Instructor Resources
Extending the Frontiers of Mathematics

Inquiries into proof and argumentation

Instructor Resources
Extending the Frontiers of Mathematics

Inquiries into proof and argumentation

Deborah J. Bergstrand

Swarthmore College

Edward B. Burger

Williams College

Deborah J. Bergstrand
Department of Mathematics
Swarthmore College
Swarthmore, PA 19081

Edward B. Burger
Department of Mathematics
Williams College
Williamstown, MA 01267

Key College Publishing was founded in 1999 as a division of Key Curriculum Press® in cooperation with Springer New York, LLC. We publish innovative texts and courseware for the undergraduate curriculum in mathematics and statistics as well as mathematics and statistic

Published by Key College Publishing, an imprint of Key Curriculum Press.

Development Editor: Kristin Burke
Editorial Production Project Manager: Laura Ryan
Text Design, Composition, and Art: Happenstance Type-O-Rama
Cover Designer: Jensen Barnes
Cover Photo Credit: Jensen Barnes/Edward B. Burger
Copyeditor: Tara Joffe
Printer: Data Reproductions Corporation
Editorial Director: Richard J. Bonacci
General Manager: Mike Simpson
Publisher: Steven Rasmussen

ISBN-13: 978-1-59757-043-5
ISBN-10: 1-59757-043-5

Printed in the United States of America
10 9 8 7 6 5 4 3 2 1   11 10 09 08 07 06

Contents

Introduction

The *Instructor Resources* contains complete solutions and commentary for each module. Introductory comments and remarks are interspersed throughout the solutions, offering background information, suggestions, or cautions about some of the challenges. Instructors may find that these comments and solutions may be useful for students, but please note that solutions are written primarily for instructors. Not all solutions include the level of detail you might expect from students.

Extending the Frontiers of Mathematics can be employed in a variety of ways. It can be the text for a course taught using an inquiry approach, in which students present their results during class meetings. It can also be used as a supplemental text for courses that either are lecture-based or have a combination of lectures and student presentations. Finally, the text can be used for independent study. In whatever context this book is used, the following features can develop and encourage mathematical thinking and curiosity:

- Puzzles and conundrums open the text, honing the student's ability to create original mathematical arguments.

- "Prove and extend *or* disprove and salvage" instruction is a recurring theme throughout, providing a consistent framework for approaching mathematics.

- Challenges of various levels of difficulty foster mathematical discovery and lively discussion.

Most challenges are presented with the instruction to "Prove and extend *or* disprove and salvage," an approach discussed at length in the preface of the student text. This instruction is used even for some challenges that are true but do not have natural extensions. The idea is to encourage students to always look for ways to extend a theorem. Praise their efforts even if the result is modest! Be clear to students about what you want presented in class, however. Though many extensions are natural, the extended version of the challenge may require a much more lengthy, abstract, or subtle proof, one that will take more class time to present and discuss. To move more quickly through material, ask students to present proofs of the valid challenges as stated in the text, with an extension stated at the end without proof. Decide what best suits your students and the goals for your course.

Planning your course

The mathematics introduced in the text covers a wide range of areas. Certainly no course can cover all the material in one semester. If adopting a pure discovery approach, then one can reasonably expect to cover roughly 10 or 11 modules (depending on the individual class, the class size, and the goals of the course).

The diagram below shows which modules depend on earlier modules. Dotted lines indicate only a modest dependence. For more detailed dependence relations, see the introductions to solutions for each module. In particular, see the introductions for dependence information on the three modules not included in the diagram: 1, 15, and 20.

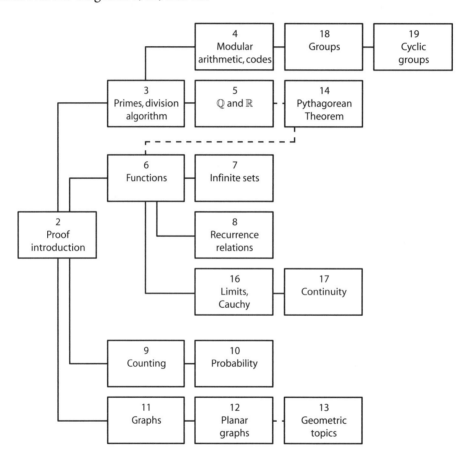

To help get the curricular juices flowing, the following list offers several different course templates and suggested modules that might be particularly appropriate, depending on the emphasis of the course. There are also optional modules that an instructor might elect to use either in their entirety or just selected portions of the material or to omit completely.

Introduction to Mathematical Proof

Modules 1, 2, 3, 4, 6, 7

Optional Modules 5, 8, 11, 13, 14, 15

Introduction to Discrete Mathematics

Modules 1, 2, 3, 4, 6, 7, 10, 11

Optional Modules 8, 12, 15

Survey of Mathematical Foundations

Modules 1, 3, 4, 5, 6, 7, 11, 14, 16, 18

Topics in Mathematics for Future Teachers

Modules 1, 2, 4, 6, 7, 10, 14, 16, 18

Optional Modules 3, 15

Capstone Senior Seminar in Mathematics

Modules 1, 6, 7, 8, 14, 16, 17, 18, 19

Optional Module 15

Commentary for instructors from author Edward Burger

This section appears as Appendix 3 in the student text.

In the opening of the student text, I remarked that the discovery approach entails uncomfortable learning. That pedagogy also leads to uncomfortable teaching. By "uncomfortable," I mean difficult and sometimes awkward. As one might expect, crafting courses with such a heavy student-discovery component is extraordinarily challenging—especially for those instructors who have not experienced inquiry-based learning in their own education.

I would argue, however, that a discovery approach to mathematics is an extremely effective means for students to wrap their minds around deep ideas, make those ideas their own, and generate new ideas. And though I confess that I do not teach all my courses in a purely discovery-based manner, the experiences I have had using this style of inquiry-based learning have resulted in some of the best and most rewarding teaching experiences in my career. Here I offer some general observations regarding this pedagogical style. In the next section I describe, in greater detail, my own experiences with using this book in the classroom.

There are many different ways to use this book with students. Obviously it could be used in any appropriate independent study work or as a supplement to any lecture-style course. In the latter case, modules would make nice projects to complement the course materials and offer a discovery based component outside of class. Alternatively, instructors might wish to adopt a hybrid instructor-lecture/ student-presentation format, in which the instructor provides lectures but also sets aside some class time for student presentations. I will focus my remarks on a pure discovery approach in which there are essentially no lectures given by the instructor; obvious adaptations can be made for lecture-style or hybrid-style courses.

A standard reality with discovery-based approaches in the classroom is that less material is covered. In my experience, I have covered about 10 or 11 modules in our 13-week semester here at Williams College. It is important to keep the class moving forward but without cutting off presenters or dismissing comments from the class. It needs to be made clear to students that if they agree to make a presentation, then they are implicitly stating that they are prepared to deliver a clear, correct, and complete solution or proof. If students get bogged down at the chalkboard, they should know that they can always back down without losing face. The instructor should be encouraging but should quickly move on to another student or "table" the challenge for another day.

If comments and questions from the audience are allowed to get out of control, then they will! The audience should understand that their role is to carefully listen to the presentation and afterward answer: *Is the presented proof correct? Complete? Clear?* Questions from the audience should always be asked in a respectful and encouraging manner (rather than in a competitive or argumentative fashion). The classroom environment should be light, friendly, and ideally festive. A common comment from an audience member is, "I see how Kate proved the result, and it looks okay to me, but I did it another way. Can I show you what I did?" Of course, pursuing this direction will bring forward progress to a halt. I avoid this scenario by stating early on that we are focusing on the proof presented. If we all agree that *that* proof is correct, complete, and clear, then we sign off on it and move on. That is, the only proof we consider is the present proof. Other proofs are fine, and even great, and should be submitted when the written work is collected. If students wish to present their (alternative) arguments, invite them to do so in your office. The class must move on.

I believe this one-proof-per-theorem format also applies (sadly) to the instructor. If a student proves a result in a correct but convoluted manner or if there is a much more elegant argument that the instructor knows, then it is tempting to show the "cool" or "correct" proof after the student presentation. Such actions could undermine the students' confidence. Showing a slicker proof might be edifying, but I would argue it is also demoralizing. Therefore I avoid sharing my favorite proofs with students (as difficult as that may be). Students should take pride in and ownership of their proofs. (I often name a result after the student who presented its proof. I've had, among many others, Son's Theorem, Nela's Theorem, and even the Matt Conjecture.)

My hope is that the challenges in this book lead to many interesting discussions between the instructor and the students. Beyond the ideas and mathematical vistas, conversations regarding issues of communicating mathematics are key. A central theme of my course is to allow students to find their own mathematical voice and produce well-written and clear arguments. Even explaining where and when we use words such as "since," "assume," "however," "because," "thus," "hence," "by," and "therefore" in the theorem-proving business is a most valuable endeavor.

In my own experience, some students often feel at the beginning of a semester as if they have been thrown into the deep end of the mathematics pool. I explicitly tell them that yes they have been thrown in, but I then assure them that their flailing to stay afloat will evolve into graceful strokes that will have them gliding through the charted and uncharted waters of mathematics. In the end, students rise up to the difficult challenge of this material, work extremely hard, feel very proud of their work, and truly appreciate their accomplishments.

How I used this material in my classroom

I developed most of this material for a one-semester course I created at Williams College in 2003. Below is the course description from the college catalog.

MATH 251: Introduction to Mathematical Proof and Argumentation

The fundamental focus of this course is for students to acquire the ability to create and clearly express mathematical arguments through an exploration of topics from discrete mathematics. Students will learn various mathematical proof techniques while discovering such areas as logic, number theory, infinity, geometry, graph theory, and probability. Our goal is not only to gain an understanding and appreciation of interesting and important areas of mathematics but also to develop and critically analyze original mathematical idea and argumentation.

Prerequisites: Calculus I or II or one year of high school calculus with permission of instructor.

The student population in the classes consisted primarily of first-and second-year students. Some students had taken linear algebra, and others had not. For most students, this was their first "proof course." I paired students into teams and determined the pairing using a Math Personality Questionnaire that I created (see Appendix). I also invited students to inform me privately if there were any students in the class they preferred not to be paired with or if they had a preference for their partner. Once students were paired up, their only allowable resources were their partners (with whom they should have met at least twice a week), the student text, and me. All other resources and individuals were off limits.

Each student was responsible for individually writing up the solution to each challenge. I asked students to type their final solutions and proofs, because that encouraged them to proofread and edit their written submissions. Every challenge was discussed in class. If students submitted a solution on or before the class in which that solution was discussed, that solution was graded out of a possible 10 points. However, if the student submitted a solution after it was discussed in class (but no more than one week later), then it was graded out of 8 points. Thus there was a penalty for not completing work before it was discussed in class, but that penalty was not fatal. Extensions to theorems did not require proofs, and students received bonus points for their generalizations.

I had an index card for each student, and I brought the stack of cards to every class. In class, I would select a student to show the group his or her solution or proof to the challenge at hand. Students could elect to pass if they had not completed the challenge, but I would note this on their index cards. For the student who offered a proof, I would note the date and the challenge and, after the presentation, note a grade from 1 to 10, depending on the quality of the student's argument, presentation, and ability to answer questions. I would select those students who had not presented recently—thus attempting to realize an even distribution of presentations among students.

The material I covered in my classes parallels that given in the opening section of this book, under the course Introduction to Mathematical Proof. Specifically, one semester, I had students explore (in this order) Modules 2, 15, 3, 4, 6, 7, 8, 9, 10, 11, 12, and a bit of 13. Another semester's syllabus was Modules 1, 2, 3, 4, 5 (first half), 6, 7, 8, 9, 10, and 11.

The final course grade was computed as follows: Class Presentations = 30%; Written Submissions = 20%; Midterm Exam = 25%; and Final Exam = 25%. The final examination was a 24-hour, open text, open notes exam. Both the midterm and final had a computational component as well as a theoretical (prove theorems) section.

I hope this commentary provides a starting point for those instructors who are considering offering their students this material in a hands-on approach. More important, I hope that it inspires instructors to craft the perfect courses for themselves and their students and that everyone enjoys the challenges and triumphs involved in discovering the wonderful world of mathematical ideas.

Puzzles and patterns
A precursor to proofs

The ten challenges in this module offer students a variety of logical puzzles to contemplate. Some challenges, especially 1.8 and 1.10, are harder than others, but all are fun. Encourage students to think about all ten with the goal of finding at least a few for which they can present clear, complete solutions. Students should not be discouraged if they can't find complete solutions. The idea is to come up with as many points of attack as possible.

Solutions

1.1. We can show that there is exactly one honest senator. The first statement guarantees at least one, call her Senator XX (not to be confused with her male colleague, Senator XY). If any other senator stands with Senator XX, the second statement tells us that at least one of the two is dishonest. We know Senator XX is honest, so it must be the other senator who is dishonest. We can make this argument for all other senators, so only Senator XX is honest.

1.2. The two statements are

   A: At least one of the boxes contains a treasured *Vass* piece.

   B: Beware: A colony of poisonous snakes awaits you in the diamond-embossed box.

We are told that either both are true or both are false. If Statement A is false, then neither box contains a *Vass* vase, so both contain snakes. But then Statement B would be true. Thus we cannot have both statements false. Therefore, both statements must be true. Statement B now tells us that there are snakes in the diamond-embossed box. Statement A tells us that at least one box contains a *Vass* piece, so it must be the box with the crossed swords.

1.3.   For the warm-up question, we can easily cover the 8×8 checkerboard with 32 dominoes: Use four horizontal dominoes for each row. To cover the board with the top two corners removed, remove one domino from the top row and slide the remaining three dominoes over appropriately. For the last challenge, look carefully at the truncated checkerboard.

We see that both deleted squares are black. This leaves 62 squares to cover with dominoes: 32 white and 30 black. Yet each domino will cover exactly one black square and one white square, so any domino arrangement will cover the same number of white squares as black squares. Thus there is no way to cover the truncated board with dominoes.

1.4.   We show that each square must be labeled with the same number. Pick an arbitrary square and suppose it is labeled n. If the four neighbor squares are not all labeled n, then at least one label must be larger and at least one must be smaller. Let the smaller label be k. Because this square has a neighbor labeled n with $n > k$, it must have another neighbor with a label still smaller than k. We can continue this process indefinitely to create an infinite, decreasing sequence. Such a sequence is impossible in the natural numbers, because the process would stop when you reach 1. Thus we must have all squares labeled identically.

1.5.   The deck contains 26 black cards and 26 red cards to begin with. If the first pile of 26 cards contains n black cards, then it contains $26 - n$ red cards. Thus the remaining n red cards must be in the other pile.

1.6.   Suppose there are k coins showing heads. Divide the coins arbitrarily into two collections, so that one collection contains k coins. The collection with k coins contains some unknown number of heads, say i, with the remaining $k - i$ coins showing tails. Note that the second collection has $k - i$ heads. Now turn over all the coins in the first collection. The i heads become tails and the $k - i$ tails become heads, which is the exact number of heads in the second collection. (Note that this challenge is closely related to the previous one.)

1.7.   One weighing is clearly insufficient; two weighings suffice. Choose three balls and weigh them against three other balls. If they balance, then the heavy ball is in the remaining trio of balls. If the scale doesn't balance, the ball will be in the heavier trio. In either case, you now have three balls to consider. Choose any two and compare them on a second scale. If they balance, the remaining ball is the heavier one. If they don't, you've once again determined the heavier ball.

1.8.   There are no Ping-Pong balls left in the barrel. Ball number 1 is removed with 1/2 minute remaining, ball number 2 is removed with 1/4 minute remaining, ball number 3 is removed with 1/8 minute remaining, and so on. So ball number n is removed with $(1/2)^n$ remaining. (Point out to dis-believers that any ball left in the barrel would have to have a number. Because you can state explicitly when each ball is removed based on its number, you have shown that the barrel is empty.) Essentially we are establishing a one-to-one correspondence between the balls and the removal times.

The Super Bonus is an entirely different matter. There is no definitive answer to the number of balls left in the barrel in this case. That is, you can remove balls in such a way as to leave as many balls in the barrel as you wish. Another way to think of this modified challenge is to imagine that the balls are still labeled 1, 2, 3, ... , but you can remove any ball you wish at each stage. If you want to leave no balls in the barrel, follow the procedure of the original challenge. If you want to leave one ball, then leave ball number 1 and remove balls 2, 3, 4, ... , in sequence, as in the original challenge. To leave any finite number of balls, say k, just start with ball number $k + 1$. To leave an infinite num-ber of balls, you could remove only even-numbered balls.

1.9.   This challenge offers a good opportunity for students to try simpler cases, moving up to find a pattern. Suggest that those who are stumped do the problem first with only two couples, then three, and so forth.

In the given challenge, Carol shook four hands. To prove this, we pair up roommates according to the number of hands they shook. Because 8-shakes shook everyone's hand except her own and her roommate's, her roommate must be 0-shakes. Now 1-shake shook only 8-shakes hand, so 7-shakes shook everyone's hand except 0-shakes, 1-shake, and his own. Thus 1-shake and 7-shakes must be roommates. Similar reasoning yields that 2-shakes and 6-shakes must be roommates, as must 3-shakes and 5-shakes. This leaves 4-shakes to be Chris's roommate, Carol.

This argument is easier to construct if you use a graph on nine vertices (the poker players minus Chris), with edges representing handshakes. (Graphs are introduced in Module 11.) The degrees of the vertices are the number of handshakes: 0, 1, 2, ... , 8. Start drawing edges from the vertex of degree 8, and the pattern becomes clear very quickly.

1.10.   This challenge is hard. Many students might discover a method that ensures half the students will get A's. One way to do this is to have the first student say the color of the hat on the second student, who then says his hat color. Then the third student says the color of the hat on the fourth student, who then says her hat color, and so on. Thus even-numbered students get A's and odd numbered students have a 50–50 chance for an A.

It is possible for the students to guarantee that all but one get A's. They agree to the following: The first student in line will look at the hats on all the remaining students and will count the number of gray hats. She will say "gray" if there is an even number of gray hats and "white" if there is an odd number. The next student can look at the remaining hats and determine his own hat color. Subse-quent students keep track of the number of gray hats behind and ahead of them and use the original parity of the number of gray hats to determine their own hat color. Thus $n - 1$ students get A's. The martyred first student has a 50–50 chance of naming her own hat color correctly.

Bringing theorems to justice
Exposing the truth through rigorous proof

This module introduces basic methods of rigorous proof in mathematics. The first five challenges focus on elementary propositional logic. Challenges 2.6 through 2.8 develop proof by contradiction or by considering cases. Challenge 2.9 requires students to use only basic axioms to prove what seems like an "obvious" result from their mathematical past.

The remainder of the module introduces proof by mathematical induction. Using a metaphor of falling dominoes, the text develops the idea of induction in detail, including two examples. Challenges 2.10 through 2.14 are classic induction proofs: All are algebraic except 2.13, which is geometric. Challenge 2.15 is substantially more difficult, offering an engaging exercise for more ambitious students.

Solutions

2.1. All are statements except for two. "Are you really taking that math course? Are you crazy?" expresses an opinion. It poses two questions and thus is not a statement. The sentence "This sentence is false" is neither true nor false. If it were true, then it would have to be false; if it were false, then it would have to be true. Thus the sentence is not a statement. While the inherent contradiction in this sentence may wow some students, the statement that really stands out as "wow" is "I prefer pi" because it is also a palindrome.

2.2. One negation for the first statement is "The number x is not negative." Note: "The number x is positive or zero" is *not* a negation because (a) it assumes that x is real (i.e., not complex), and (b) it requires a theorem that asserts that every real number falls into one of the three following disjoint collections: positive numbers, negative numbers, or zero.

Here are sample negations for the remaining statements: "There exists an x for which $f(x)$ is not less than 0" and "For every x, $g'(x)$ is defined."

2.3. "Not (A and B)" is equivalent to "(Not A) or (Not B)."

A	B	A and B	Not (A and B)	Not A	Not B	(Not A) or (Not B)
T	T	T	F	F	F	F
T	F	F	T	F	T	T
F	T	F	T	T	F	T
F	F	F	T	T	T	T

"Not (A or B)" is equivalent to "(Not A) and (Not B)."

A	B	A and B	Not (A and B)	Not A	Not B	(Not A) and (Not B)
T	T	T	F	F	F	F
T	F	T	F	F	T	F
F	T	T	F	T	F	F
F	F	F	T	T	T	T

Remark: The *or* in mathematics is an *inclusive or* because the statement "A or B" is considered true when *both* A and B are true as well as when just one of A or B is true and the other is false. An *exclusive or* would preclude both A and B being true. The exclusive or is often implied in common speech, such as "You may have ice cream or cookies for dessert," much to the dismay of young children.

2.4. All the implications are true except "If $\sin \pi = 0$, then $\cos \pi = 0$." In this statement, the hypothesis is true, but the conclusion is false; therefore, the implication is false.

Remark: Students often struggle with the truth value of implications. Remind them that knowing an implication is true does not guarantee the hypothesis or conclusion is true. Rather, a true implication excludes *only* the case that the hypothesis is true while the conclusion is false. An implication such as "If $1 + 1 = 3$, then $\sqrt{36} = -6$" is sometimes called *vacuously true* because the hypothesis will never be satisfied.

2.5. An implication and its contrapositive are equivalent. An implication's converse and inverse are equivalent.

A	B	A implies B	Not A	Not B	(Not B) implies (Not A)
T	T	T	F	F	T
T	F	F	F	T	F
F	T	T	T	F	T
F	F	T	T	T	T

A	B	B implies A	Not A	Not B	(Not A) implies (Not B)
T	T	T	F	F	T
T	F	T	F	T	T
F	T	F	T	F	F
F	F	T	T	T	T

2.6. The statement is a theorem. *Extensions:* The sum of an even number of odd numbers is an even number; the sum of an even number of even numbers is even; the sum of an odd number of odd numbers is odd; the sum of an even number and an odd number is odd. (There are more!)

Proof: We prove the first claim. Proofs of the rest are analogous. Let $n_1, n_2, \ldots, n_k$ be odd numbers, with k even. Because k is even, we have $k = 2j$ for some integer j. Each n_i is not even and, therefore, must equal $2m_i + 1$ for some integer m_i. Thus we have

$$n_1 + n_2 + \cdots + n_k = 2m_1 + 1 + 2m_2 + 1 + \cdots + 2m_k + 1$$
$$= 2m_1 + 2m_2 + \cdots + 2m_k + k$$
$$= 2(m_1 + m_2 + \cdots + m_k) + 2j = 2(m_1 + m_2 + \cdots + m_k + j),$$

which is clearly even.

2.7. The statement is a theorem. *Extension:* The product of any n consecutive numbers is a multiple of n.

Proof: Consider the n consecutive numbers $k, k+1, \ldots, k+(n-1)$. It suffices to show that one of these numbers is a multiple of n. But there are at most $n-1$ distinct integers between any two consecutive multiples of n, so the set $k, k+1, \ldots, k+(n-1)$ must contain a multiple of n. The result follows.

2.8. The statement is false. *Counterexample:* Consider the four integers 92, 93, 95, and 96. *Salvage:* If the average of four distinct integers is 94, then at least one of the integers must be greater than or equal to 96.

Proof: Assume all four integers are less than 96. Then their sum is at most $95 + 94 + 93 + 92$, so their average is at most 93.5. This contradicts the given hypothesis; therefore, our assumption must be wrong, and at least one number is 96 or greater.

2.9. *Proof:* If a and b are two arbitrary positive integers, then $-a$ and $-b$ are two arbitrary negative integers. We will show that $(-a)(-b) = ab$. Because a and b are positive and ab is just b added to itself a times, we know that ab is positive, and we'll be done.

First observe that $(-a)(-b) = ((-1)a)((-1)b) = (-1)(-1)ab$, by the commutativity and associativity of multiplication. So now we need only show that $(-1)(-1) = 1$. We know that $1 + (-1) = 0$, so we have $(-1)(1 + (-1)) = 0$, by a given property of 0. By the distributive law, we have $(-1)(1) + (-1)(-1)$ $= -1 + (-1)(-1) = 0$, using 1 as the multiplicative identity.

Now add 1 to both sides of this last equation and use the associative law of addition to get $1 + (-1) + (-1)(-1) = 1$, which reduces to $(-1)(-1) = 1$, which was what we wanted.

Remark: Students may construct quite elaborate proofs of this result. Focus on whether they use only the given axioms rather than efficiency.

2.10. *Proof:* We will use induction to show that for any integer $n \geq 1$,

$$1 + 3 + 5 + 7 + \cdots + (2n - 1) = n^2.$$

First we establish the base case, when $n = 1$. This requires that $1 = 1^2$, which is true. Next we show that if the condition holds for some arbitrary value $k \geq 1$, then it must hold for $k + 1$. We assume that $1 + 3 + 5 + 7 + \cdots + (2k - 1) = k^2$. We must show that $1 + 3 + 5 + 7 + \cdots + (2k - 1) + (2(k + 1) - 1)$ $= (k + 1)^2$. By our induction hypothesis, the left side of the expression is equal to $k^2 + (2(k + 1) - 1)$. This simplifies to $k^2 + 2k + 1 = (k + 1)^2$, as desired.

2.11. *Proof:* We use induction on n. For $n = 1$, the left side becomes $1^2 = 1$ and the right side becomes $\frac{1(1+1)(3)}{6} = 1$, so the base case holds. Now assume the result holds for $n = k \geq 1$:

$$1^2 + 2^2 + 3^2 + \cdots + k^2 = \frac{k(k+1)(2k+1)}{6}.$$

Adding $(k + 1)^2$ to both sides, we obtain

$$1^2 + 2^2 + 3^2 + \cdots + k^2 + (k+1)^2 = \frac{k(k+1)(2k+1)}{6} + (k+1)^2$$

$$= \frac{2k^3 + 3k^2 + k}{6} + k^2 + 2k + 1$$

$$= \frac{2k^3 + 9k^2 + 13k + 6}{6} = \frac{(k+1)(k+2)(2k+3)}{6},$$

as desired.

2.12.   *Proof:* We use induction on n. The base case $n = 2$ holds because

$$\left(1-\frac{1}{4}\right)=\frac{3}{4}=\frac{n+1}{2n}.$$

Assume that for $n = k$,

$$\prod_{i=2}^{k}\left(1-\frac{1}{i^2}\right)=\frac{k+1}{2k}.$$

Multiplying both sides by $(1-\frac{1}{(k+1)^2})$ gives

$$\prod_{i=2}^{k+1}\left(1-\frac{1}{i^2}\right)=\frac{k+1}{2k}\left(1-\frac{1}{(k+1)^2}\right)=\frac{k+1}{2k}\left(\frac{k^2+2k}{(k+1)^2}\right)$$
$$=\frac{(k+1)(k+2)k}{(k+1)(k+1)(2k)}=\frac{k+2}{2k+2}=\frac{(k+1)+1}{2(k+1)}.$$

Thus it follows from the principle of induction that the first statement is true for all $n \ge 2$.

The formal identity is obtained as follows:

$$\lim_{n\to\infty}\prod_{n=2}^{n}\left(1-\frac{1}{n^2}\right)=\lim_{n\to\infty}\frac{n+1}{2n}=\lim_{n\to\infty}\frac{1}{2}+\frac{1}{2n}=\frac{1}{2}.$$

2.13.   The statement is a theorem.

Proof: We use induction on the number of lines, n. Given one line, we choose one side of the line to be gold and the other side to be purple. Assume that for n lines, there exists a coloring of the regions in the plane. Now suppose we have $n + 1$ lines in the plane. We momentarily remove one, say line ℓ, so we have n lines in the plane. Thus by the inductive hypothesis, we can color the plane so that each region can be painted either purple or gold in such a manner that any two regions sharing a common border edge will be painted different colors. Now place line ℓ back into our picture. Flip all the colors on one side of ℓ. Now on either side of ℓ, we still have a valid coloring. For any region intersected by line ℓ, we have switched the color on one side of ℓ such that the old region is now two regions colored different colors. Thus we have a valid coloring for all regions created using $n + 1$ lines. So by mathematical induction, the result holds for all $n \ge 1$.

2.14.   The statement is a theorem.

Proof: We use induction on n. When $n = 0$, the statement becomes $(2)(1) - 1 = 1^0$, which clearly holds. Now suppose the statement holds for n: $F_{n+2}F_n - (F_{n+1})^2 = (-1)^n$.

We need to prove the statement for $n + 1$:

$$F_{n+3}F_{n+1} - (F_{n+2})^2 = (-1)^{n+1}.$$

The definition of the Fibonacci numbers allows us to replace F_{n+3} with $F_{n+1} + F_{n+2}$, so the left side becomes

$$(F_{n+1} + F_{n+2})F_{n+1} - (F_{n+2})^2 = (F_{n+1})^2 + F_{n+1}F_{n+2} - (F_{n+2})^2.$$

We can also replace (F_{n+2}) with $F_n + F_{n+1}$ and reduce further to

$$(F_{n+1})^2 + F_{n+1}(F_n + F_{n+1}) - (F_n + F_{n+1})^2 = 2(F_{n+1})^2 + F_nF_{n+1} - ((F_n)^2 + 2F_nF_{n+1} + (F_{n+1})^2)$$

$$= (F_{n+1})^2 - F_nF_{n+1} - (F_n)^2 = (F_{n+1})^2 - (F_nF_{n+1} + (F_n)^2)$$

$$= (F_{n+1})^2 - (F_n(F_n + F_{n+1})).$$

Now we reintroduce an F_{n+2} and factor out -1:

$$(F_{n+1})^2 - F_nF_{n+2} = -(F_{n+2}F_n - (F_{n+1})^2).$$

By our induction hypothesis, the right side reduces to $-(-1)^n = (-1)^{n+1}$. Thus our proof is complete.

Remark: When trying to prove an identity, students may be tempted to start with the desired equation and manipulate both sides until they obtain some obvious equality, such as $1 = 1$. They then may claim that their original identity must hold. Point out that this approach is valid only if each step is reversible. Encourage them to work on the left side alone until they obtain the right side, or vice versa.

2.15.   The statement is a theorem.

Proof: We use induction on n. The result holds for $n = 0$ as follows:

$$\frac{p_n}{q_n} = \frac{a_0}{1} = a_0 \, .$$

Assume that, for any two sequences defined as in the challenge, the result holds for n:

$$\frac{p_n}{q_n} = a_0 + \cfrac{1}{a_1 + \cfrac{1}{\ddots \, a_{n-1} + \cfrac{1}{a_n}}}.$$

We want to show the identity holds for $n + 1$:

$$\frac{p_{n+1}}{q_{n+1}} = a_0 + \cfrac{1}{a_1 + \cfrac{1}{\ddots \, a_{n-1} + \cfrac{1}{a_n + \cfrac{1}{a_{n+1}}}}}.$$

Define a new set of positive integers $A_i = a_i$, for $0 \leq i \leq n-1$, and $A_n = a_n + \frac{1}{a_{n+1}}$. Define numbers r_i and s_i analogous to p_i and q_i but using the integers $A_0, A_1, \ldots, A_n$. Note that for $i < n$, $r_i = p_i$ and $s_i = q_i$.

Then by our induction hypothesis, we have

$$\frac{r_n}{s_n} = A_0 + \cfrac{1}{A_1 + \cfrac{1}{\ddots\, A_{n-1} + \cfrac{1}{A_n}}} = a_0 + \cfrac{1}{a_1 + \cfrac{1}{\ddots\, a_{n-1} + \cfrac{1}{A_n}}} = a_0 + \cfrac{1}{a_1 + \cfrac{1}{\ddots\, a_{n-1} + \cfrac{1}{a_n + \cfrac{1}{a_{n+1}}}}}.$$

By definition, $r_n = A_n r_{n-1} + r_{n-2} = (a_n + \frac{1}{a_{n+1}})p_{n-1} + p_{n-2}$.

So $r_n = a_n p_{n-1} + p_{n-2} + \frac{1}{a_{n+1}} p_{n-1} = p_n + \frac{1}{a_{n+1}} p_{n-1}$.

Rewriting, we obtain $a_{n+1} r_n = a_{n+1} p_n + p_{n-1} = p_{n+1}$. Thus $r_n = \frac{p_{n+1}}{a_{n+1}}$. An analogous argument produces $s_n = \frac{q_{n+1}}{a_{n+1}}$. Therefore $\frac{r_n}{s_n} = \frac{p_{n+1}}{q_{n+1}}$, and our result holds by induction.

Stepping back

Proof by induction is a good method to use if the property can be described systematically for an arbitrary nonnegative integer. Otherwise try proof by contradiction. Assume the set of numbers with the given property is finite. Then you know there is a largest or last number, which often leads to a contradiction.

3

Delving into the dependable digits
Counting on counting numbers

Module 3 introduces elementary number theory, including the Division Algorithm, the Euclidean Algorithm, and basic results about prime numbers. The material offers students an excellent opportunity to refine the proof-writing skills introduced in Module 2. The most profound and elegant result in the module is Challenge 3.15: Euclid's result that there are infinitely many primes. This material works particularly well for an introductory course on proofs or higher mathematics.

Solutions

3.1. If we factor the numbers, we see that $168 = 2^3 \cdot 3 \cdot 7$ and $180 = 2^2 \cdot 3^2 \cdot 5$. Therefore it is clear that the $\gcd(168, 180) = 12$. Because factoring large numbers is difficult, we would not want to compute the $\gcd(3913, 23177)$ in the same way. Thus the answer to the second question is "No."

Proof of useful fact: Suppose a, b, d, m, and n are integers such that $d\,|\,m$ and $d\,|\,n$. Then there exist integers k and l such that $m = kd$ and $n = ld$. Thus $am + bn = akd + bld = (ak + bl)d$, where $ak + bl$ is an integer. Therefore $d\,|\,(am + bn)$.

3.2. *Proof:* Let $g = \gcd(m, n)$ and $g' = \gcd(m, r)$. Because $n = mq + r$, any divisor of m and r is a divisor of n. Also, because $r = n - mq$, any divisor of n and m is a divisor of r. Thus because g divides m and n, it also divides r. So we have $g\,|\,\gcd(m, r)$—in other words, $g\,|\,g'$. By similar reasoning, because g' divides m and r, we have g' dividing n. Hence $g'\,|\,\gcd(m, n)$. In other words, $g'\,|\,g$. Therefore there exist integers l and l' such that $g' = lg$ and $g = l'g'$. Thus we see $g' = ll'g'$, and so $ll' = 1$. Because g and g' are both positive, we have that $l = l' = 1$, which implies that $g = g'$.

3.3. The statement is a theorem. *Extension:* Given natural numbers m, n, and k, there exists a unique pair of integers q and r such that $n = mq + r$ and $k \le r < (m + k)$.

Proof of original statement: The integer multiples of m partition the real line into intervals of length m. Any integer n must either equal a unique multiple of m or lie in a unique interval between two consecutive multiples. That is, given any integer n, there is a unique integer q such that

$mq \le n < m(q + 1)$. Let $r = n - mq$. Then r is clearly unique and satisfies $0 \le r < m$. Thus $n = mq + r$ and $0 \le r < m$.

Proof of extension: Note that when we add k to each multiple of m, we get another partition of the real line into intervals of length m. So the given integer n must either equal $mq + k$ for a unique q or lie between two consecutive endpoints: $mq + k < n < m(q + 1) + k$. Thus we again have unique integers q and r such that $n = mq + r$, except now $k \le r < m + k$.

Remark: The notation here is suggestive, of course. When we divide n by m, we get an integer *quotient q* and a *remainder r*. More significantly, the proofs for Challenges 3.3, 3.4, and 3.5 implicitly use the Well-Ordering Principle.

3.4.   *Proof of the Division Algorithm:* We follow the same approach as for Challenge 3.3. The integer multiples of m partition the real line; the integer n lies in a unique interval or on its left endpoint. So we have a unique q and r for which $n = mq + r$. Because r must be nonnegative, we have $0 \le r < |m|$.

3.5.   *Proof of the Euclidean Algorithm:* We need to prove two things: the process terminates after a finite number of steps, and the individual statements hold. The first line and all subsequent lines follow by repeated applications of the Division Algorithm. Because we are generating a strictly decreasing sequence of nonnegative integers, $r_1, r_2, \ldots, r_L$, there are only finitely many steps possible within the bounds of the natural numbers. Note again the use of the Well-Ordering Principle. Thus the process terminates with remainder $r_L = 0$.

Now apply Challenge 3.2 successively to note that $\gcd(m, n) = \gcd(m, r_1) = \gcd(r_1, r_2) = \cdots = \gcd(r_{L-2}, r_{L-1})$. Because $r_L = 0$, then $\gcd(r_{L-2}, r_{L-1}) = r_{L-1}$, and the result holds.

3.6.   We apply the Euclidean Algorithm to obtain the following:

$$23177 = 3913(5) + 3612$$
$$3913 = 3612(1) + 301$$
$$3612 = 301(12) + 0.$$

Thus the $\gcd(3913, 23177) = 301$.

Remark: Note that the Euclidean Algorithm makes this much easier than we might have anticipated back in Challenge 3.1.

3.7.   In Challenge 3.6, take each equation generated by the Euclidean Algorithm, except the last one, and solve for the remainder. Starting at the top, substitute the quantity for r_1 given by the first equation into the second equation. Then substitute the result into the next equation. Repeat until you reach the last equation. First we obtain all the remainders:

$$3612 = 23177 - 3913(5)$$
$$301 = 3913 - 3612(1).$$

And then

$$301 = 3913 - (23177 - 3913(5))(1).$$

Simplifying, we obtain

$$301 = 3913(6) + 23177(-1).$$

Thus we find $x = 6$ and $y = -1$.

Remark: Students may realize that this technique applies to any pair of integers m and n. You can always find integers x and y so that $mx + ny = \gcd(m, n)$. This result is a special case of the salvaged version of Challenge 3.8, below.

3.8. The statement is false. *Counterexample:* $12(2) + 18(2) = 60$, but the $\gcd(12, 18) = 6$, not 60. *Salvage:* Let n, m, and g be integers. There exist integers x and y satisfying the linear Diophantine equation $mx + ny = g$ if and only if $\gcd(m, n) \mid g$.

Proof: We can easily generalize the method from Challenge 3.7 to show that solving for the remainders in the Euclidian Algorithm will always give us integer solutions x and y that satisfy $mx + ny = \gcd(m, n)$. Now assume the $\gcd(m, n) \mid g$. Say $\gcd(m, n) = gk$ for some k. Multiply $mx + ny = \gcd(m, n)$ by k to obtain the result. Conversely, assume there is a solution to $mx + ny = g$. The $\gcd(m, n)$ divides the left side, so it must divide the right.

Remark: If your students are inspired by this result, encourage them to look for an important extension, which is actually a generalization of Challenge 3.9, below. Let n, m, and g be integers. If $\gcd(m, n) \mid g$, then there are infinitely many pairs of integers x and y satisfying the linear Diophantine equation $mx + ny = g$. Moreover, if (x_0, y_0) is a solution, then (x, y) is a solution if and only if $x = x_0 + nt$ and $y = y_0 - mt$ for some integer t. The proof of this result is analogous to that given below for Challenge 3.9.

3.9. The statement is a theorem. *Extension:* If (x_0, y_0) is a solution to the linear Diophantine equation, then all solutions to $mx + ny = 1$ can be expressed as (x_t, y_t), where $x_t = x_0 + nt$ and $y_t = y_0 - mt$ for $t \in \mathbb{Z}$.

Proof: As stated in the proof of the salvage of Challenge 3.8, we can always find integer solutions x_0 and y_0 so that $mx_0 + ny_0 = \gcd(m, n)$. Thus m and n are relatively prime if and only if there exist integers x_0 and y_0 so that $mx_0 + ny_0 = 1$. To prove the extension, we substitute the given x_t and y_t to obtain $mx_t + ny_t = m(x_0 + nt) + y(y_0 - mt) = mx_0 + mnt + ny_0 - mnt = 1$.

3.10. (a) The statement is false. *Counterexample:* Let $k = 6$, $m = 3$, and $n = 4$. *Salvage:* Let k, m, and n be integers. If $k \mid mn$ and k and m are relatively prime, then $k \mid n$.

Proof: Given that k and m are relatively prime, Challenge 3.9 implies that there exist integers x and y such that $mx + ky = 1$. Thus we have $mnx + kny = n$. Given that $k \mid mn$, we have k dividing both terms on the left side of the equation; therefore, $k \mid n$.

3.10. (b)   The statement is a theorem. *Extension:* If $a_1, a_2, \dots, a_n$ are pairwise relatively prime and $a_i \mid m$ for $i = 1, 2, \dots, n$, then $a_1 a_2 \cdots a_n \mid m$.

Proof of original statement: First note that the result holds if one or both of a and b is ± 1. So we assume that neither a nor b is ± 1. We have $a \mid m$ and $b \mid m$, so $m = ak = bl$ for some integers k and l. Thus $a \mid bl$ and $a \nmid b$; otherwise we would have $\gcd(a, b) = a$. So, by part (a), we have $a \mid l$. Therefore $l = aj$ for some integer j, which implies $m = bl = baj$. Therefore $ab \mid m$.

Proof of extension: We use induction on n. The base case for $n = 2$ follows from part (a). Suppose the result holds for $n = k$ pairwise relatively prime integers. Consider $n = k + 1$ pairwise relatively prime integers $a_1, a_2, \dots, a_k, a_{k+1}$, with the property that $a_i \mid m$ for $i = 1, 2, \dots, k + 1$. Look first at just a_1 and a_2. By part (a), we know that $a_1 a_2 \mid m$. We also claim that $a_1 a_2$ shares no common factors with any of $a_3, a_4, \dots, a_{k+1}$. Suppose $b \mid a_i$. Given that a_i and a_1 are relatively prime, we can have neither b nor any factor of b dividing a_1; similarly for a_2. Thus no factor of b divides $a_1 a_2$, which implies $a_1 a_2$ is relatively prime to a_i. Now we have k pairwise relatively prime integers $a_1 a_2, a_3, \dots, a_{k+1}$, each dividing m. So by our induction assumption, we must have $a_1 a_2 \cdots a_{k+1} \mid m$. Thus our results hold by mathematical induction.

3.11.   The statement is a theorem. There is no obvious extension expected here.

Proof: Let $p > 5$ be prime. Thus p is odd and not divisible by 3. Observe that $p^2 - 1 = (p - 1)(p + 1)$, which is the product of two even numbers, because p is odd. Thus $4 \mid (p^2 - 1)$. We also note that every third integer is divisible by 3. Thus, because 3 does not divide p, we must have 3 dividing either $p - 1$ or $p + 1$. Therefore $12 \mid (p^2 - 1)$.

3.12.   The statement is a theorem. *Extension:* Let $m_1, m_2, \dots, m_k$ be any finite set of integers. If p is a prime such that $p \mid m_1 m_2 \dots m_k$, then $p \mid m_i$ for some i, $1 \le i \le k$.

Proof: First we prove the original statement. Assume $p \mid mn$. If $p \mid m$, we're done. So assume p doesn't divide m. Because p is prime, we have that p and m are relatively prime. By the salvaged version of Challenge 3.10, we must have that $p \mid n$, and we're done.

Now suppose $p \mid m_1 m_2 \dots m_k$. By the proof above, if p does not divide m_k, then $p \mid m_1 m_2 \dots m_{k-1}$. Apply the result above repeatedly to obtain the generalization. (Formalize using mathematical induction.)

3.13.   *Proof of the Fundamental Theorem of Arithmetic:* Let $n > 1$ be an integer. First we show that n can be written as a product of primes. If n is prime, we are done. If not, then n can be written as the product of two smaller positive integers, both greater than 1. If these are both primes, we are done. Otherwise, we continue this process until n is written as a product of primes. (We know the process will end, because the factors are getting smaller at each step yet must always remain greater than 1.) To formalize this proof, we would need Strong Mathematical Induction. Note also the use of the Well-Ordering Principle.

Now we show that this product is unique up to reordering of the factors. Suppose we have

$$n = p_1 p_2 \cdots p_k = q_1 q_2 \cdots q_m,$$

where all the p's and q's are primes. Thus p_1 divides the product of the q_j's. So by Challenge 3.12, we have that p_1 divides some q_j. Because both are prime, we have $p_1 = q_j$. Divide both sides by this prime and repeat the argument. Eventually each p_i matches up with a q_j and vice versa.

3.14. The statement is a theorem. *Extension:* If m and n are integers with $m > 1$ and $m \mid n$, then $m \nmid (n + k)$, where $0 < k < m$.

Proof of extension: Suppose $m \mid (n + k)$ for some k with $0 < k < m$. Then because $m \mid n$, we must also have $m \mid k$, which is impossible if $0 < k < m$.

3.15. *Euclid's proof of the infinity of primes:* Suppose there are only finitely many primes, $p_1, p_2, \ldots,$ p_n, for some positive integer n. Now let $N = p_1 p_2 \cdots p_n$. All assumed primes divide N, and each prime is greater than 1. Therefore by Challenge 3.14, no prime divides $N + 1$. But this contradicts the Fundamental Theorem of Arithmetic. So there must be infinitely many primes.

Stepping back

Let C be the set of all even positive integers. Then any integer equal to twice an odd number will be prime in C. This leads to many nonunique factorizations, such as $36 = 6 \times 6 = 2 \times 18$.

4

Going around in circles

The art of modular arithmetic

Module 4 introduces the basic ideas and results on congruences, including the Chinese Remainder Theorem (4.7) and Fermat's Little Theorem (4.10). Challenges 4.1 through 4.12 offer students a substantial glimpse of this elegant area of elementary number theory. The remaining material presents a powerful application—public key cryptography—with substantial contextual detail. This module relies heavily on material from Module 3. It is a wonderful addition to an introductory course on proof, foundations of math, or math for teachers.

Solutions

4.1. The three properties hold.

Proof: Let a, b, c, and m be integers, with $m > 1$. We have $a \equiv a \bmod m$ for all a because $m \mid (a - a)$. Thus congruence is reflexive.

   If $a \equiv b \bmod m$, then $m \mid (a - b)$, and so $m \mid (b - a)$. Thus $b \equiv a \bmod m$, and we have that congruence is symmetric.

   Finally, if $a \equiv b \bmod m$ and $b \equiv c \bmod m$, then $m \mid (a - b)$ and $m \mid (b - c)$. Thus $m \mid (a - b + b - c)$, giving $m \mid (a - c)$. So we have that congruence is transitive.

Remark: Note that the definition of *congruence modulo m* requires that m be greater than 1. If students wonder why, ask them to think about what *congruence mod* 1 would mean. Once they discover that all integers are congruent mod 1, they'll realize that requiring $m > 1$ avoids this trivial case.

4.2. The statement is a theorem. *Extension:* Let a, b, and m be integers with $m > 1$. Then $a \equiv b \bmod m$ if and only if $u = v$ whenever we express $a = im + u$ and $b = jm + v$ for integers i, j, u, and v, with $-m < u - v < m$.

Proof of original statement: First suppose a and b have the same remainder when divided by m. Then there exist integers i, j, and r so that $a = im + r$ and $b = jm + r$. Thus $a - b = (i - j)m$; so, $m \mid (a - b)$, and we have $a \equiv b \bmod m$. For the converse, suppose $a \equiv b \bmod m$. Thus $m \mid (a - b)$, and we have

$a - b = nm$ for some integer n. By the Division Algorithm, we have integers i, j, r, and s, so that $a = im + r$ and $b = jm + s$, with $0 \leq r, s < m$. Thus $a - b = (i - j)m + r - s$. Setting these two expressions for $a - b$ equal, we obtain $(i - j)m + r - s = nm$, which implies $r - s = (n - i + j)m$, a multiple of m. But because $0 \leq r, s < m$, this multiple must be 0; so we must have $r = s$.

Proof of extension: First suppose $a \equiv b \bmod m$ and suppose we have $a = im + u$ and $b = jm + v$ for integers i, j, u, and v, with $-m < u - v < m$. Then m divides $a - b$, and so m divides $(im + u) - (jm + v)$. Then m divides $(i - j)m + (u - v)$, which implies m divides $u - v$. But because we have $-m < u - v < m$, we must have $u - v = 0$ and therefore $u = v$. For the converse, use the Division Algorithm to obtain integers i, j, r, and s so that $a = im + r$ and $b = jm + s$, with $0 \leq r, s < m$. Then we also know that $-m < r - s < m$. So by the hypothesis, we have that in this case we can conclude $r = s$. Therefore $a \equiv b \bmod m$.

4.3.   The statement is a theorem. *Extension:* Let S be a set of m integers. Suppose that when each element of S is divided by m, the remainders obtained are all distinct. Then given any integer a, there exists a unique $r \in S$ such that $a \equiv r \bmod m$.

Proof of original statement: Given a and m, by the Division Algorithm (Challenge 3.4), we know there exist unique integers q and r such that $a = qm + r$, with $0 \leq r < m$. Thus $a - r = qm$, so $m \mid (a - r)$, and we have $a \equiv r \bmod m$. Moreover, $0 \leq r < m$, so $r \in S$.

To establish uniqueness, suppose there exist r_1 and r_2, $0 \leq r_1, r_2 \leq m - 1$, such that $a \equiv r_1 \bmod m$ and $a \equiv r_2 \bmod m$. Then $a - r_1$ and $a - r_2$ are multiples of m, so there exist integers q_1 and q_2 such that $a - r_1 = q_1 m$ and $a - r_2 = q_2 m$. Thus we have $a = q_1 m + r_1$ and $a = q_2 m + r_2$, with $0 \leq r_1, r_2 \leq m - 1$. But by the uniqueness result of the Division Algorithm, we must have $r_1 = r_2$. Therefore the challenge holds.

Proof of extension: The extension follows easily from the original result. The distinct remainders derived from S are precisely the elements of the set $R = \{0, 1, 2, \ldots, m - 1\}$. Given a, we know from the proof above that there is a unique $k \in R$ with $a \equiv k \bmod m$. Let r be the element of S that has remainder $k \bmod m$; the extended result follows.

4.4.   The statement is a theorem. *Extension:* This result also holds for more than two summands or factors.

Proof of original statement: Given $a \equiv b \bmod m$, we have $a - b = qm$ for some q. Similarly $c - c' = q'm$ for some q'. Adding the two equations, we get $(a + c) - (b + c') = (q + q')m$, which implies that $a + c \equiv b + c' \bmod m$. Moreover, we have $ac - bc = qcm$ and $bc - bc' = bq'm$. Adding these two equations yields $ac - bc' = (qc + bq')m$. Therefore $ac \equiv bc' \bmod m$.

4.5.   The statement is false. *Counterexample:* Let $a = 3$, $b = 6$, $m = 6$, and $c = 4$. Then $ac = 12$ and $bc = 24$, which are both 0 mod 6, but 3 and 6 are not equivalent mod 6. *Salvage:* Let a, b, c, and m be integers with $m > 1$. Then $a \equiv b \bmod m$ implies $ac \equiv bc \bmod m$. For a partial converse, if c and m are relatively prime, then $ac \equiv bc \bmod m$ implies $a \equiv b \bmod m$.

Proof: Let a, b, c, and m be as given, with $a \equiv b \bmod m$. Then $m \mid (a - b)$, which implies that $m \mid (a - b)c$. Thus we have $m \mid (ac - bc)$; so $ac \equiv bc \bmod m$.

For the partial converse, suppose a, b, c, and m are as given, with c and m relatively prime. Then by the spectacularly useful Challenge 3.9, there exist integers x and y satisfying the linear Diophantine equation $cx + my = 1$. Thus multiplying through by a, we have $acx + amy = a$; multiplying through by b, we have $bcx + bmy = b$. Subtracting these two identities and factoring their difference, we have $(ac - bc)x + (ay - by)m = a - b$. If $ac \equiv bc \bmod m$, then $m \mid (ac - bc)$. Thus m divides both factored expressions on the left side of the identity. Therefore m must divide the right side, $a - b$. Thus we have $a \equiv b \bmod m$.

Remark: The moral of this proof is you can never overemphasize the value of Challenge 3.9. When presented with integers a and b that are relatively prime, the first thing to remember is that the equation $ax + by = 1$ has integer solutions x and y.

Note also that Challenges 4.4 and 4.5 are valuable tools for reducing arithmetic expressions mod m, provided the conditions of relative primality are met. Challenge 4.5 is critical in the proof of Fermat's Little Theorem (Challenge 4.10). As a further example, let $m = 3$. Because m is prime, it is relatively prime to any other integer. So by the salvage of Challenge 4.5, we have $11x \equiv 2x \equiv -x \bmod 3$ for any integer x. We also have $4y \equiv y \bmod 3$ for any integer y. Therefore, by the salvage of Challenge 4.5, we also have $11x + 4y \equiv -x + y \bmod 3$. This process is important in Challenge 4.12 and in the discussion of public key cryptography at the end of the module.

4.6. Given integers a, b, and m, with $m > 1$, the equation $ax \equiv b \bmod m$ has a solution if and only if $\gcd(a, m) \mid b$.

Proof: The equation $ax \equiv b \bmod m$ has a solution if and only if there is an integer x such that $ax - b$ is a multiple of m. This occurs if and only if there is an integer y such that $ax - b = my$, which occurs if and only if $ax + m(-y) = b$. But by the salvage for Challenge 3.8, this occurs if and only if $\gcd(a, m) \mid b$.

4.7. The statement is a theorem (the Chinese Remainder Theorem). *Extension:* Let $m_1, m_2, \ldots, m_k$ be integers greater than 1 that are pairwise relatively prime. Then for integers a and b, the system of simultaneous linear congruences

$$ax \equiv b \bmod m_1$$

$$ax \equiv b \bmod m_2$$

$$\vdots$$

$$ax \equiv b \bmod m_k$$

has a solution $x \in \mathbb{Z}$ if and only if there is an integer solution to the linear congruence

$$ax \equiv b \bmod m_1 m_2 \cdots m_k.$$

Proof of original statement: Suppose first that the system of linear congruences has a solution x. Then there exist integers y_1 and y_2 such that $ax - b = m_1 y_1$ and $ax - b = m_2 y_2$. Therefore $m_1 y_1 = m_2 y_2$, so $m_2 \mid m_1 y_1$. Because m_1 and m_2 are relatively prime, by the salvage of Challenge 3.10, we have $m_2 \mid y_1$. Thus $y_1 = m_2 y_3$ for some integer y_3. Substituting into the first linear congruence, we obtain $ax - b = m_1 m_2 y_3$. Therefore $ax \equiv b \bmod m_1 m_2$.

Now suppose the congruence $ax \equiv b \bmod m_1 m_2$ has a solution x. Then there exists an integer y such that $ax - b = m_1 m_2 y$. But then $ax - b = m_1(m_2 y)$ and $ax - b = m_2(m_1 y)$, so x is also a solution to the individual congruences.

Proof of extension: We use induction. The proof above covers the base case. Suppose the result holds for k integers $m_1, m_2, \ldots, m_k$ greater than 1 that are pairwise relatively prime. Consider a set of $k + 1$ such integers $m_1, m_2, \ldots, m_{k+1}$. From the base case, we know that the system composed of the first two congruences $ax \equiv b \bmod m_1$ and $ax \equiv b \bmod m_2$ has a solution x if and only if $ax \equiv b \bmod m_1 m_2$. So in our set of $k + 1$ integers, replace m_1 and m_2 with the single integer $m_1 m_2$. The resulting set of k integers will still be pairwise relatively prime. By induction, we also know that the set of k linear congruences

$$ax \equiv b \bmod m_1 m_2$$
$$ax \equiv b \bmod m_3$$
$$\vdots$$
$$ax \equiv b \bmod m_{k+1}$$

has a solution $x \in \mathbb{Z}$ if and only if there is an integer solution to the linear congruence

$$ax \equiv b \bmod m_1 m_2 \ldots m_{k+1}.$$

And the result holds.

Remark: This result is called the Chinese Remainder Theorem because it was known in China in the first century AD.

4.8. The statement is a theorem. *Extension:* Let a, k, and m be integers, with a and m relatively prime and $m > 1$. Then $(k + 1)a, (k + 2)a, \ldots, (k + m - 1)a$ are distinct mod m.

Proof of original statement: Suppose $ra \equiv sa \bmod m$, with $1 \leq r, s \leq m - 1$. Because a and m are relatively prime, $r \equiv s \bmod m$, by the salvage of Challenge 4.5. Thus $m \mid (s - r)$, so $s - r$ is a multiple of m. By considering the largest and smallest possible values for r and s, we also know that $-m + 2 \leq s - r \leq m - 2$. Clearly the only multiple of m that lies between $-m + 2$ and $m - 2$ inclusive is 0. Therefore $r = s$, and the result follows.

Proof of extension: Suppose $(k + r)a \equiv (k + s)a \bmod m$, with $1 \leq r, s \leq m - 1$. Then $ra \equiv sa \bmod m$, with $1 \leq r, s \leq m - 1$. By the original statement, we have $r = s$, and the result holds.

4.9. When you reduce the elements of the second set mod p, you obtain exactly the elements of the first set. This follows from Challenge 4.8.

4.10. *Proof:* Consider the sets $\{1, 2, \ldots, p-1\}$ and $\{a, 2a, \ldots, (p-1)a\}$. By Challenge 4.9, these sets are congruent mod p. Thus the products of their respective elements are also congruent mod p. So we have $a^{p-1}(p-1)! \equiv (p-1)! \bmod p$. Because p is prime, it must be relatively prime to $(p-1)!$; so by the salvage to Challenge 4.5, we have $a^{p-1} \equiv 1 \bmod p$.

4.11. The statement is false. *Counterexample:* Let $a = 3$ and $m = 4$. Then $3^4 = 81$, which is not congruent to 3 mod 4. *Salvage:* Let a and p be two integers with p prime. Then $a^p \equiv a \bmod p$.

Proof: If p is prime and $p \mid a$, then $p \mid a^p$, so the result holds. If $p \nmid a$, then by Fermat's Little Theorem (Challenge 4.10), we have $a^{p-1} \equiv 1 \bmod p$, which implies $a^p \equiv p \bmod p$.

4.12. *Proof:* We will show that both 3 and 5 divide $11n^8 + 4n^4$. Then because 3 and 5 are relatively prime, the Chinese Remainder Theorem (Challenge 4.7) will give the result. Because we want to reduce $11n^8 + 4n^4$ mod 3 and then mod 5, we first make some observations. Challenge 4.4 implies that we can reduce each term separately. The salvage to Challenge 4.5 implies that we can reduce the factors in each term separately, because both 3 and 5 are prime and hence are relatively prime to any power of n and to any coefficient. Thus we have

$$11n^8 + 4n^4 \equiv -n^8 + n^4 \bmod 3 \qquad \text{(reducing the coefficients mod 3)}$$
$$\equiv -(n^3(n^3)n^2) + n^4 \bmod 3 \qquad \text{(factoring)}$$
$$\equiv -nnn^2 + n^4 \bmod 3 \qquad \text{(salvage of Challenge 4.11)}$$
$$\equiv -n^4 + n^4 \equiv 0 \bmod 3.$$

So 3 divides $11n^8 + 4n^4$. Now we do the same reduction mod 5:

$$11n^8 + 4n^4 \equiv n^8 - n^4 \bmod 5 \qquad \text{(reducing the coefficients mod 5)}$$
$$\equiv n^5 n^3 - n^4 \bmod 5 \qquad \text{(factoring)}$$
$$\equiv nn^3 - n^4 \bmod 5 \qquad \text{(salvage of Challenge 4.11)}$$
$$\equiv n^4 - n^4 \equiv 0 \bmod 5.$$

So 5 divides $11n^8 + 4n^4$. Thus we have $11n^8 + 4n^4 \equiv 0 \bmod 3$ and $11n^8 + 4n^4 \equiv 0 \bmod 5$. So by the Chinese Remainder Theorem (Challenge 4.7), we have $11n^8 + 4n^4 \equiv 0 \bmod 15$. Therefore 15 divides $11n^8 + 4n^4$ for any integer n.

4.13. *Proof:* Because e and m are relatively prime, Challenge 3.9 implies there are integers x_0 and y_0 such that $ex_0 + my_0 = 1$. The extension to Challenge 3.9 implies that for any integer t, $x = x_0 + mt$ and $y = y_0 - et$ also satisfy $ex + my = 1$. We can choose t so that $x_0 + mt$ is positive and then let d be this value of $x_0 + mt$. Then $ed = 1 - my$, which we can rewrite as $ed = 1 + m(-y)$, and the result holds.

4.14.   *Proof:* We are given that p and q are distinct primes, $n = pq$, $m = (p-1)(q-1)$, and $e > 0$ is an integer relatively prime to m. We also have integers $d > 0$ and y such that $ed = 1 + my$. Finally we have an integer message W, $1 \leq W < n$, that has been encoded into C, satisfying $W^e \equiv C \bmod n$ and $1 \leq C < n$. To decode C, we find D, $1 \leq D < n$, such that $C^d \equiv D \bmod n$. We must show that $D = W$.

Observe that $D \equiv C^d \equiv (W^e)^d \bmod n$ by repeated applications of the extension to Challenge 4.4. Now we will show that $W^{ed} \equiv W \bmod n$. We first note that $W^{ed} = W^{1+my} = W \cdot W^{my} = W \cdot W^{(p-1)(q-1)y}$. Now observe that because $1 \leq W < n = pq$, we cannot have W divisible by both p and q.

If $p \nmid W$, then $W^{p-1} \equiv 1 \bmod p$ by Fermat's Little Theorem, which implies that $W \cdot W^{(p-1)(q-1)y} \equiv W(1)^{(q-1)y} \equiv W \bmod p$. Similarly, if $q \nmid W$, we have $W \cdot W^{(p-1)(q-1)y} \equiv W(1)^{(p-1)y} \equiv W \bmod q$. Then by the Chinese Remainder Theorem, we have $W \cdot W^{(p-1)(q-1)y} \equiv W \bmod pq$. So $W^{ed} \equiv W \bmod n$.

Now recall that both W and D are positive integers less than n. Because they are in the same congruence class mod n, we must have $D = W$.

Remark: Note that this result holds for any d such that $ed = 1 + my$. We will need this fact when we discuss breaking a public key code in *Stepping back*.

4.15.   Everyone knows the public part of your code—namely, the number e used for encoding messages they send to you. No one knows your private decoding number d. Here's a signature that only you could create. First, transform your name into a number N in some straightforward way, with $1 \leq N < n$. Let N^d be your signature message S. When a recipient, say Alice, receives a message from you with your signature S attached, she computes the remainder of S^e when divided by n. She finds $S^e \equiv N^{de} \equiv N \bmod n$, and she recognizes your name, thus authenticating the message.

Note that Alice now has your authenticating signature S, so she could use it to send messages in your name. Thus this method of authentification really only works once. This is one reason public key code users frequently change their primes.

Stepping back

One way to break someone's public key code is to determine the secret decoding number d. If you could factor n into pq, then you would know $m = (p-1)(q-1)$. You already know e, the public encoding number. Because e and m are relatively prime, Challenge 3.9 implies there are integers x_0 and y_0 such that $ex_0 + my_0 = 1$. To find x_0 and y_0, you follow the example of Challenge 3.7. Then by the proof of Challenge 4.13, you know $d = x_0 + mt$, where t is chosen so that $d > 0$. Any such d will work, as noted in the remark following the proof of Challenge 4.14. Therefore you can decode a transmitted code word C by finding D, $1 \leq D < n$, such that $C^d \equiv D \bmod n$; the code is broken.

The time it would take to apply this code-breaking method is basically the time it would take to factor n. Public key coding is effective because factoring very large numbers is extremely difficult. However, there may be some other approach that is much less time-consuming. Whether such a method exists is an open research question. If someone finds an efficient code-breaking algorithm, the current system of public key cryptography will be rendered obsolete.

5

The irrational side of numbers
A world of nonrepeating digits

This module presents one of the classic proofs in mathematics: the irrationality of $\sqrt{2}$. The first four challenges develop and extend this elegant proof by contradiction and are a must in any course on mathematical proof. The remaining challenges offer more results on irrational numbers, including a proof that e is irrational, and would be excellent material for a capstone course in a math major or a course on math for teachers.

Solutions

5.1. The answer to the question is "Yes." *Theorem:* If a and b are nonzero integers, the number $\frac{a}{b} + \frac{b}{a}$ is an integer if and only if $a = b$.

Proof: If $a = b$, then $\frac{a}{b} + \frac{b}{a} = 2$, which is an integer. For the converse, suppose $\frac{a}{b} + \frac{b}{a}$ is an integer. Then $\frac{a}{b} + \frac{b}{a} = \frac{a^2 + b^2}{ab}$ is an integer, so $ab \mid (a^2 + b^2)$. Now suppose, in addition, that $a \neq b$. By the Fundamental Theorem of Arithmetic (Challenge 3.13), we know that a and b can be factored into products of primes. If $a \neq b$, then there is a prime p such that $p \mid a$ but $p \nmid b$. In particular, note that this implies $p \nmid b^2$. But we also have $p \mid ab$ and $p \mid a^2$, which implies $p \mid b^2$, a contradiction. Therefore $a = b$.

Remark: After this challenge, the text informally introduces the real numbers, $\mathbb{R}$, as the collection of points in the number line. Many students may already realize that $\mathbb{R}$ is, in fact, a difficult set to define formally. There are two classic methods: Dedekind cuts and the completion of the set of Cauchy sequences of rational numbers. Reassure students that they will see $\mathbb{R}$ formally defined in a course on real analysis.

5.2. The statement is false. *Counterexample:* Consider $m = n = 0$. *Salvage:* There do not exist nonzero integers m and n satisfying $m^2 = 2n^2$.

Proof: Suppose m and n are nonzero integers such that $m^2 = 2n^2$. Then $2 \mid m^2$. Because 2 is prime, Challenge 3.12 implies that $2 \mid m$, so $m = 2k$ for some integer k. Now we have $(2k)^2 = 2n^2$. So $4k^2 = 2n^2$, which simplifies to $2k^2 = n^2$. But now we have $2 \mid n$. Repeating the argument, we will get that $2 \mid k$. Our logic can be reapplied any finite number of times, but we cannot divide the integer m by 2 an infinite number of times. Therefore our original assumption was false, and the result holds.

Remark: This result holds for all primes, not just 2. If some students prove such an extension, that's great, but note that it also comes up in Challenge 5.4.

5.3.   *Proof:* Suppose $\sqrt{2}$ is rational. Then $\sqrt{2} = \frac{m}{n}$ for some integers m and n, $n \neq 0$. Thus $2 = \frac{m^2}{n^2}$, and so $m^2 = 2n^2$. But we know from the previous challenge that this is impossible for n nonzero. Therefore $\sqrt{2}$ is irrational.

5.4.   *Proof:* Suppose $\sqrt{p}$ is rational. Then $\sqrt{p} = \frac{m}{n}$ for some integers m and n, $n \neq 0$. Thus $p = \frac{m^2}{n^2}$, and so $m^2 = pn^2$. We will follow the same approach as for the salvage of Challenge 5.2, noting first that we have $p \mid m^2$. Because p is prime, Challenge 3.12 implies that $p \mid m$. Thus $m = pk$ for some integer k. Note that k must be less than m because $p \geq 2$. Now we have $(pk)^2 = pn^2$, so $p^2k^2 = pn^2$, which simplifies to $pk^2 = n^2$. But now we have $p \mid n$. Repeating the argument, we get that $p \mid k$. Our logic can be reapplied, but we cannot divide the integer m by p an infinite number of times. Therefore our original assumption was false, and the result holds.

 We don't reach a contradiction for $\sqrt{4}$. Here's where the argument breaks down. From the point $m^2 = 4n^2$, we deduce that $4 \mid m^2$, but then we cannot conclude that $4 \mid m$. Our only valid conclusion is that $2 \mid m$, which gives us $m = 2k$. Thus $m^2 = 4n^2$ becomes $4k^2 = 4n^2$, which implies $k^2 = n^2$. We have no contradiction, nowhere to go. This is reassuring, of course, because $\sqrt{4}$ is not at all irrational. Also, there do exist nonzero integers m and n for which $m^2 = 4n^2$. For example, $m = 6$ and $n = 3$.

5.5.   *Proof:* Suppose $m > 1$ is an integer with $\sqrt{m}$ rational. Then $\sqrt{m} = \frac{r}{s}$ for some integers r and s, $s \neq 0$. We want to show that $s = 1$. Without loss of generality, we may assume that r and s are positive and relatively prime. Thus we also have r^2 and s^2 relatively prime. Having $\sqrt{m} = \frac{r}{s}$ gives us $m = \frac{r^2}{s^2}$, which must be in lowest terms. Because m is an integer, we must have $s^2 = 1$. Because s is positive, we must have $s = 1$. Therefore $\sqrt{m}$ is an integer.

 Finally, we note that $5 = \sqrt{25} < \sqrt{35} < \sqrt{36} = 6$. Therefore $\sqrt{35}$ is not an integer. Thus we have that $\sqrt{35}$ is not rational.

Remark: Note that this argument also establishes that the prime factors of m occur in pairs, and thus m is a perfect square, as the result implies.

5.6.   The statement is a theorem. There are no immediate extensions.

Proof: From the previous challenge, we need only show that $\sqrt{F_{n+2}F_n}$ is not an integer—that is, $F_{n+2}F_n$ is not a perfect square. We know from Challenge 2.14 that $F_{n+2}F_n - (F_{n+1})^2 = (-1)^n$. Therefore $F_{n+2}F_n = (F_{n+1})^2 + (-1)^n$. Because $n > 1$, $(F_{n+1})^2$ is a perfect square greater than 3. Thus we need only show that perfect squares do not appear as consecutive integers. Observe that if n is a positive integer, then the next largest perfect square after n^2 is $(n+1)^2$. But $(n+1)^2 = n^2 + 2n + 1$, which is

$2n + 1 > 1$ larger than n^2. Thus $(F_{n+1})^2 + (-1)^n$ cannot be a perfect square; so $\sqrt{F_{n+2}F_n}$ is not an integer. Again, by the contrapositive of the previous challenge, we have $\sqrt{F_{n+2}F_n}$ irrational.

Follow-up: The result does not hold for $\sqrt{F_n}$. When $n = 11$, $F_{11} = 144$; so $\sqrt{F_{11}} = 12$.

5.7. The statement is false. *Counterexample:* Consider the number $\sqrt{2}^{\sqrt{2}}$. If this number is rational, then we have a counterexample with $\tau_1 = \tau_2 = \sqrt{2}$. If the number is irrational, then we have a counterexample with $\tau_1 = \sqrt{2}^{\sqrt{2}}$ and $\tau_2 = \sqrt{2}$.

Remark: The number $\sqrt{2}^{\sqrt{2}}$ is actually irrational, though this is very difficult to prove. In fact, $\sqrt{2}^{\sqrt{2}}$ is transcendental. A *transcendental number* is a complex number that is not the zero of a nontrivial polynomial with integer coefficients. Complex numbers that are the zeros of such nontrivial polynomials are called *algebraic numbers.* So all integers are algebraic, as are all rational numbers. For example, $\sqrt{2}$ is algebraic because it is a zero of the polynomial $x^2 - 2$.

The fact that $\sqrt{2}^{\sqrt{2}}$ is transcendental follows from a result proven independently by Gelfond and Schneider in 1934: If α and β are algebraic numbers with $\alpha \neq 0$, 1 and β irrational, then α^β is transcendental. This theorem answered the seventh of the 23 famous open questions posed by leading mathematician David Hilbert in 1900 at the International Mathematical Congress in Paris.

5.8. *Proof:* Clearly the sum is positive. Now note the following inequality:

$$\frac{1}{2} + \frac{1}{2 \cdot 3} + \frac{1}{2 \cdot 3 \cdot 4} + \cdots < \frac{1}{2} + \frac{1}{2 \cdot 2} + \frac{1}{2 \cdot 2 \cdot 2} + \cdots,$$

the right side of which is a geometric series having sum $\frac{1/2}{1 - 1/2} = 1$. The result follows.

5.9. *Proof:* The series expansion for e equals 2 plus the sum shown to be less than 1 in the previous challenge. Thus $2 < e < 3$, and e is not an integer.

5.10. *Proof:* Following the hint in Appendix 1, suppose that e is a rational number r/s. We may assume that both r and s are positive. By the previous challenge, e is not an integer, and so $s \geq 2$. So now we have

$$\frac{r}{s} = 1 + \frac{1}{1!} + \frac{1}{2!} + \frac{1}{3!} + \cdots + \frac{1}{n!} + \cdots.$$

Multiplying both sides by $s!$ yields

$$r(s-1)! = s! + s! + s(s-1)\cdots(4)(3) + \cdots + s + 1 + \frac{1}{s+1} + \frac{1}{(s+1)(s+2)} + \cdots.$$

The left side of this expression is an integer, as is the sum of integer terms on the right side, so the sum of the fractions must also be an integer. Certainly the fractions have a positive sum. Now apply the same reasoning used in Challenge 5.8 to obtain

$$\frac{1}{s+1} + \frac{1}{(s+1)(s+2)} + \frac{1}{(s+1)(s+2)(s+3)} + \cdots < \frac{1}{s+1} + \frac{1}{(s+1)^2} + \frac{1}{(s+1)^3} + \cdots.$$

So again we have a geometric series, and its sum is $\frac{1/(s+1)}{1-1/(s+1)} = \frac{1}{s+1-1} = 1/s \le 1/2$. Thus the sum of fractions is a positive integer less than 1/2, which yields our contradiction. Thus e is irrational.

Remark: This beautiful result is due to 18th-century Swiss mathematician Leonard Euler. The elegant proof above was discovered in 1815 by French mathematician Joseph Fourier.

Stepping back

Claim: A real number is rational if and only if its decimal expansion terminates or repeats. (It is hoped that most students have seen these conditions at some point in their math education. It is still an interesting exercise for them to construct a rigorous argument proving that the conditions are necessary and sufficient.)

Proof: Suppose x is a rational number. Then $x = r/s$ for integers r and s, with $s \ne 0$. To derive the decimal expansion of x, use long division to divide s into r. At each step of the process, the intermediate differences obtained will always lie between 0 and s. If at any point we get a difference of 0, then our decimal expansion terminates. If not, then because there are only a finite number of possible differences, at some step to the right of the decimal point, a difference must repeat. Thereafter the decimal expansion will repeat. (This argument foreshadows the pigeonhole principle coming in Module 9.)

To prove the converse, suppose x is a real number with a decimal expansion that terminates or repeats. If the expansion terminates, then x can be written as a quotient of integers where the denominator is an appropriate power of 10. If the decimal expansion repeats with a period of k digits for some integer k, then the decimal number $10^k x - x = x(10^k - 1)$ will terminate and, thus, can be written as r/s for integers r and s. This gives $x = r/s(10^k - 1)$, and we have that x is rational.

Claim: The number $n.999\ldots$ is rational for any $n \in \mathbb{N}$. Moreover, $n.999\cdots = n + 1$.

Proof: We will show the result holds for $n = 1$. The argument is analogous for other values of n. Let $x = 1.999\ldots$. Then $10x = 19.999\ldots$, so we have $10x - x = 18$. Thus $9x = 18$, so $x = 2$. Clearly x is rational.

Claim: The number $x = \sum_{n=1}^{\infty} 10^{-n^2}$ is irrational.

Proof: Expanding x yields $x = 10^{-1} + 10^{-4} + 10^{-9} + \cdots$. Therefore $x = 0.1001000010000001\ldots$. In this expansion, clearly the number of zeros between two consecutive ones is always increasing. Therefore x does not have a repeating or terminating decimal expansion and is therefore irrational.

Now for the real question: What is the likelihood that a real number chosen at random is irrational? The answer is 1. To put it another way, the probability that a real number chosen at random is rational is 0. We can actually prove a stronger claim: The probability that a real number chosen at random from the unit interval is rational is 0. (Note that the following proof relies on limits and probability.)

Claim: If x is a real number, $0 \leq x \leq 1$, chosen at random, the probability that x is rational is 0.

Proof: There are many ways to specify how a number might be chosen at random. We will generate digits d_i to the right of the decimal point independently and randomly from among 0, 1, … , 9. This produces a number $x = 0.d_1 d_2 d_3 \ldots$. For each decimal place of x, each of the digits 0, 1, … , 9 appears with probability 1/10. (The value of each digit could be determined by rolling a fair, 10-sided die.)

Now suppose x is rational. Then we know that the decimal expansion must end in a repeating pattern, possibly a string of all zeros. So there is a positive integer k such that a specific pattern of k digits repeats forever after some point in the list $0.d_1 d_2 d_3 \ldots$. Thus after the first occurrence of the pattern, every digit that follows is completely determined. Because the digits of x are generated at random, each of these predetermined digits would appear with probability 1/10. Because the digits are produced independently, the probability of a particular sequence of digits is the product of the probabilities of each digit. So the probability that all the remaining digits of x really do match the pattern must be $\lim_{n \to \infty}(1/10)^n$, which equals 0. (We take a limit as n goes to infinity, because there are infinitely many digits remaining in the decimal representation of x.) Therefore the probability that x is rational is 0.

Remark: Note that this result also establishes that a real number selected at random from anywhere on the number line will be rational with probability 0. This mirrors a fact from analysis that the rational numbers have Lebesgue measure 0.

6

Discovering how to function in mathematics

Moving beyond ordinary relations

This module begins with the basic definitions of set theory and functions and then develops the tools needed for some very sophisticated techniques and results. Students may be familiar with much of the notation and terminology in the early challenges. The task for these students is to reconcile their previous function experience with the notation and approach used here, as well as to produce clear and concise arguments.

Following basic set theory, one-to-one and onto functions are introduced, leading naturally to inverse functions. These ideas are fundamental throughout mathematics and are covered in Challenges 6.1 through 6.7. Thereafter, results on the cardinality of finite sets and the power set lay the groundwork for material in Module 7 on infinity. The module ends with the very challenging Schroeder-Bernstein Theorem (Challenge 6.14), which may stump even your most talented students, so be cautious about assigning it. The special case offered in Challenge 6.15 is subtle but much more accessible. Neither of these two results is a prerequisite for further work. They might be used as interesting discussion points rather than assigned for formal proofs.

Solutions

6.1. $\emptyset \in A$ is false. *Counterexample:* $A = \{1, 2, 3\}$ does not have $\emptyset$ as an element. *Salvage:* $\emptyset \subseteq A$.

Proof: We need to show that every element of $\emptyset$ is an element of A. This holds vacuously because $\emptyset$ has no elements.

$A \subseteq A$ is true. *Proof:* Every element of A is clearly an element of A.

$A \cap B \subseteq A \cup B$ is true. *Proof:* If $x \in A \cap B$, then by definition of intersection, $x \in A$. So by the definition of union, $x \in A \cup B$. The result follows.

$(A \setminus B) \setminus C = A \setminus (B \setminus C)$ is false. *Counterexample:* Let $A = \{x, y, z\}$, $B = \{y, z\}$, $C = \{z\}$. Then $(A \setminus B) \setminus C = \{x\}$, but $A \setminus (B \setminus C) = \{x, z\}$. *Salvage:* $(A \setminus B) \setminus C \subseteq A \setminus (B \setminus C)$. *Proof:* If $x \in (A \setminus B) \setminus C$, then x lies in A but not in B or C. Therefore x is not an element of $B \setminus C$, and so it would remain in $A \setminus (B \setminus C)$.

$A \cup (B \cap C) = (A \cap B) \cup (A \cap C)$ is false. *Counterexample:* Let A be a set with an element x that is in neither B nor C. Then x lies in $A \cup (B \cap C)$ but not in $(A \cap B) \cup (A \cap C)$. *Salvage:* $A \cup (B \cap C) = (A \cup B) \cap (A \cup C)$. *Proof:* We will show these two sets are equal by showing that x is in one set if and only if it is in the other set. Now $x \in (A \cup B) \cap (A \cup C)$ if and only if $x \in A \cup B$ and $x \in A \cup C$. This holds if and only if x lies in A or B and x lies in A or C. But this holds if and only if $x \in A$ or $x \in B \cap C$—that is, if and only if $x \in A \cup (B \cap C)$. Thus the two sets are equal.

Remark: Note that this result establishes a distributive law for union over intersection. The companion law holds as well: $A \cap (B \cup C) = (A \cap B) \cup (A \cap C)$ for all sets A, B, and C. The proof is analogous.

$A \setminus B \subseteq A$ is true. *Proof:* If $x \in A \setminus B$, then x lies in A but not in B. Therefore $x \in A$, and the result holds.

$A \subseteq B$ if and only if $A \cup B = B$ is true. *Proof:* Suppose $A \subseteq B$. Then every element of A also lies in B. So every element of $A \cup B$ must lie in B. But every element of B also lies in $A \cup B$, which implies $A \cup B = B$. Now suppose $A \cup B = B$ and consider $x \in A$. Then x lies in $A \cup B$ and so also in B. Thus $A \subseteq B$, and our result holds.

6.2.   The statement is false. *Counterexample:* Look at the function $f = \{(a, 1), (b, 1)\}$. (Note that this example is given in the text just prior to this challenge.) *Salvage:* Let $f: X \to Y$ be a function, and suppose that $x_0 \in X$ and $y_0 \in Y$ satisfy $(x_0, y_0) \in f$. If $(x, y) \in f$ and $y \neq y_0$, then $x \neq x_0$.

Proof: Suppose that $x = x_0$. By the definition of a function, with $(x_0, y_0) \in f$ and $(x, y) \in f$, we must have $y = y_0$. We have shown the contrapositive, and therefore the result holds.

Remark: The requirement that each x in the domain of a function corresponds to exactly one y in the codomain can be confusing for students. The expression *exactly one* is often misinterpreted to mean that a y-value can be used only once as the function value for an x. This, however, is the requirement for a *one-to-one* function, which is introduced prior to Challenge 6.5. If this question comes up, emphasize that saying each element in the domain corresponds to *exactly one* element in the codomain merely means that each domain element has *one* function value—not more, not fewer.

6.3. $|f(X)| \leq |Y|$ and $|f(X)| \leq |X|$.

Proof: We know $f(X)$ is a subset of Y, and so it can have no more elements than Y. Similarly, because each element of X has exactly one image in $f(X)$, the set $f(X)$ has at most as many elements as X.

6.4.   The statement is a theorem. *Extension:* Let $n > 1$ be an integer. If $f_1 : X_1 \to X_2, f_2 : X_2 \to X_3,$ $\dots, f_n : X_n \to X_{n+1}$ are functions, then the composition $f_n \circ f_{n-1} \circ \cdots \circ f_1$ is a function. In particular, $f_n \circ f_{n-1} \circ \cdots \circ f_1 : X_1 \to X_{n+1}$, so $(f_n \circ f_{n-1} \circ \cdots \circ f_1)(x) = f_n(f_{n-1}(\dots (x)))$. (See the *Remark* that follows for comments about associativity.)

Proof: We give the proof for the original statement about g and f. The more general result follows by induction. To show $g \circ f$ is a function from X to Z, consider $x \in X$. Because f is a function from X to Y, there is exactly one $y \in Y$ so that $(x, y) \in X \times Y$. Because g is a function from Y to Z and we have $y \in Y$, there is exactly one $z \in Z$ so that $(y, z) \in Y \times Z$. Therefore by the definition of $g \circ f$, we have $(x, z) \in g \circ f$. Furthermore, by the derivation of z, we have that for each $x \in X$, there is exactly one $z \in Z$ with the property that $(x, z) \in g \circ f$. Thus $g \circ f$ is a function from X to Z.

Remark: For $f_1 \circ f_2 \circ \cdots \circ f_n$ to be well defined as defined in the extension, composition of functions must be associative. So for $f: X \to Y$, $g: Y \to Z$, and $h: Z \to W$, we need to show that $h \circ (g \circ f) = (h \circ g) \circ f$. For the composition on each side, (x, w) lies in the function if and only if for each $x \in X$ there exists exactly one $y \in Y$ and exactly one $z \in Z$ such that $(x, y) \in f$, $(y, z) \in g$, and $(z, w) \in h$. These conditions are satisfied because f, g, and h are functions.

6.5.   There are many possible examples, of course. We give one for each case.

- ***f* is one-to-one and onto.** The simplest example here is the identity function $f(x) = x$, but the proof that it is one-to-one and onto is too trivial to be instructive. So consider instead the function $f(x) = 3x + 1$. To show f is one-to-one, suppose $f(x_1) = f(x_2)$. Then $3x_1 + 1 = 3x_2 + 1$, which simplifies to $x_1 = x_2$. To show f is onto, consider an arbitrary $y_0 \in \mathbb{R}$, the codomain. We need to show there is an $x_0 \in \mathbb{R}$, the domain, for which $f(x_0) = y_0$. Thus we solve $y_0 = 3x_0 + 1$ for x_0 to obtain $x_0 = \frac{y_0 - 1}{3}$, which does lie in the domain. Thus $f(\frac{y_0 - 1}{3}) = y_0$, and f is onto.

Remark: Students will usually find one-to-one proofs straightforward, especially if they learn to begin their proofs with something like "Suppose $f(x_1) = f(x_2)$." The proof then often follows nicely either by manipulating the equation or by applying other properties of the given function. Proving a function is onto, however, is often more difficult. Students can get confused with the "work backward from what you want" approach, even though this is often the best thing to do. Encourage them to begin their proofs with something like "Consider an arbitrary element y_0 in the codomain," followed by something like "We need to find an x_0 in the domain for which $f(x_0) = y_0$." Finding x_0 may involve solving the equation, or it may require applying other properties of the function, but at least students have a place to start.

- ***f* is one-to-one but not onto.** Let f be the function $f(x) = e^x$. Then f is one-to-one, because whenever $e^{x_1} = e^{x_2}$, we take the natural log of both sides to obtain $x_1 = x_2$. Also, f is not onto because for any $y \le 0$ in the codomain, there is no x in the domain for which $f(x) = y$.

- ***f* is onto but not one-to-one.** Let f be the function $f(x) = x(x - 1)(x + 1)$. Then f is not one-to-one because, for example, $f(0) = f(1) = 0$, but $0 \ne 1$. Also, f is an odd degree cubic polynomial, the graph of which clearly shows it to be onto.

- ***f* is neither one-to-one nor onto.** Let f be the function $f(x) = x^2$. Then f is not one-to-one because, for example, $f(3) = f(-3) = 9$, but $3 \ne -3$. Also, f is not onto because for any $y < 0$ in the codomain, there is no x in the domain for which $f(x) = y$.

6.6.   The answer to both questions is "Yes."

Proof: Suppose $f: X \to Y$ and $g: Y \to Z$ are functions. Suppose further that both are one-to-one. To show $g \circ f$ is one-to-one, suppose $g(f(x_1)) = g(f(x_2))$ for some x_1 and x_2 in X. Then g one-to-one implies $f(x_1) = f(x_2)$, and f one-to-one implies $x_1 = x_2$. Assume now that f and g are both onto. To show $g \circ f$ is onto, consider $z_0 \in Z$. Then g onto implies there exists a $y_0 \in Y$ such that $g(y_0) = z_0$. Furthermore, f onto implies there exists an $x_0 \in X$ such that $f(x_0) = y_0$. Thus we have $g(f(x_0)) = z_0$, showing $g \circ f$ is onto.

6.7.   The statement is false. *Counterexample:* Let $f: \mathbb{R} \to \mathbb{R}$ be given by $f(x) = x^2$. Then f^{-1} contains, for example, the pairs $(4, 2)$ and $(4, -2)$, which violates the definition of a function. *Salvage:* Given a function $f: X \to Y$, its inverse $f^{-1}: f(X) \to X$ is also a function if and only if f is one-to-one. Moreover, for all $y \in f(X)$, $(f \circ f^{-1})(y) = y$, and for all $x \in X$, $(f^{-1} \circ f)(x) = x$.

Proof: Suppose $f: X \to Y$ is one-to-one. Then by definition, whenever (x_1, y) and (x_2, y) are in f, we must have $x_1 = x_2$. Therefore for each $y \in f(X)$, there is exactly one $x \in X$ such that $(y, x) \in f^{-1}$. Thus f^{-1} is a function from $f(X)$ to X.

   To show that for all $x \in X$, $(f^{-1} \circ f)(x) = x$, suppose $(x, y) \in f$. Then by the definition of f^{-1}, we have $(y, x) \in f^{-1}$, and by the definition of $f^{-1} \circ f$, we have $(x, x) \in f^{-1} \circ f$. So $(f^{-1} \circ f)(x) = x$. Similarly, for all $y \in f(X)$, $(f \circ f^{-1})(y) = y$.

Remark: Note that the inverse function f^{-1} is defined with domain equal to $f(X)$, not Y. For $f: X \to Y$ to have an inverse function $f^{-1}: Y \to X$, we must have $f(X) = Y$. That is, the function f must be onto as well as one-to-one.

   Beware also that the notation f^{-1} is standard but tricky. There is the usual mistake for those students who see the "-1" as an exponent and thus confuse f^{-1} with $1/f$. In addition, note that this text uses f^{-1} to denote a set of ordered pairs that may not define a function. Finally, though the definition of *preimage* that follows Challenge 6.7 is standard, it gives rise to another source of confusion for students. Here's an example: Given a function $f: X \to Y$ and an element $y_0 \in Y$, the preimage of y_0 is the set of all elements $x \in X$ that map to y_0 under the function f. This preimage is denoted $f^{-1}(\{y_0\})$. Emphasize to students that the expression $f^{-1}(\{y_0\})$ may thus be a set of values, not just a single function value, unless f^{-1} is indeed a function. Moreover, if y_0 does not lie in the image of f, the preimage $f^{-1}(\{y_0\})$ will be the empty set.

6.8.   The statement is false. *Counterexample:* The function $f: \mathbb{N} \to \mathbb{N}$ given by $f(n) = 2n$ is one-to-one but not onto. (See the *Remark* below.) *Salvage:* If X and Y are finite sets, a one-to-one function $f: X \to Y$ is onto if and only if $|X| = |Y|$.

Proof: Suppose X and Y are finite and $f: X \to Y$ is one-to-one and onto. Because f is a function, we know that each $x \in X$ appears in exactly one pair (x, y) in f. Because f is one-to-one, we know that each $y \in Y$ appears in at most one pair (x, y) in f. Because f is onto, we know that each y appears in at least one pair (x, y). Thus each y appears in exactly one pair as well. Thus we have a one-to-one correspondence between elements of X and elements of Y, which proves that $|X| = |Y|$.

Now suppose X and Y are finite, $|X| = |Y|$, and $f : X \rightarrow Y$ is one-to-one. Because f is one-to-one, we know that no elements of X can share function values; thus $|f(X)| = |X|$, which equals $|Y|$. Therefore every element of Y must be used as the image of some element of X, and we have that f is onto.

Remark: We do not yet have a definition of the cardinality of an infinite set, but the salvaged result above foreshadows a definition given in Module 7. Note that the salvage of Challenge 6.10 states that finite sets A and B have the same cardinality if there is a one-to-one, onto function from one to the other. In Module 7, this definition is stated for all sets, infinite as well as finite.

6.9.   There is such a function. The function $f : \mathbb{Z} \rightarrow 2\mathbb{Z}$ given by $f(x) = 2x$ is both one-to-one and onto.

Proof: To show f is one-to-one, suppose $f(x_1) = f(x_2)$. Then $2x_1 = 2x_2$, and so $x_1 = x_2$. To show f is onto, consider $y \in 2\mathbb{Z}$. Then y must equal $2x$ for some integer x, and therefore $y = f(x)$.

6.10.   The statement is false. *Counterexample:* The function $f : \{1, 2\} \rightarrow \{1, 2, 3\}$ given by $f(x) = x$ is one-to-one, but the domain and codomain do not have the same cardinality. *Salvage:* Two finite sets A and B have the same cardinality if and only if there exists a one-to-one, onto function $f : A \rightarrow B$.

Proof: Suppose $f : A \rightarrow B$ is a one-to-one, onto function with A and B finite sets. Then by the proof for Challenge 6.8, we have $|A| = |B|$. Now suppose A and B are finite, with $|A| = |B|$. Then we can put the elements of A into one-to-one correspondence with the elements of B. This correspondence will give us a set of pairs (a,b) that we can use to define a function $f : A \rightarrow B$. Because every $b \in B$ appears in a pair, f is onto. Because no two elements of A correspond to the same element in B, f is one-to-one.

6.11.   The elements of $\mathcal{P}(\{\clubsuit, \diamondsuit, \heartsuit, \spadesuit\})$ are $\emptyset$, $\{\clubsuit\}$, $\{\diamondsuit\}$, $\{\heartsuit\}$, $\{\spadesuit\}$, $\{\clubsuit, \diamondsuit\}$, $\{\clubsuit, \heartsuit\}$, $\{\clubsuit, \spadesuit\}$, $\{\diamondsuit, \heartsuit\}$, $\{\diamondsuit, \spadesuit\}$, $\{\heartsuit, \spadesuit\}$, $\{\clubsuit, \diamondsuit, \heartsuit\}$, $\{\clubsuit, \diamondsuit, \spadesuit\}$, $\{\clubsuit, \heartsuit, \spadesuit\}$, $\{\diamondsuit, \heartsuit, \spadesuit\}$, $\{\clubsuit, \diamondsuit, \heartsuit, \spadesuit\}$.

6.12.   Because $|A| = n$, we can write $A = \{a_1, a_2, \dots , a_n\}$. A function $f : \mathcal{P}(A) \rightarrow \{0, 1\}^n$ has as its domain the set of all subsets of A. So to define f, we need to associate with each subset of A an ordered n-tuple of 0's and 1's. Let B be a subset of A. We define $f(B)$ as follows: $f(B) = (t_1, t_2, \dots , t_n)$, where

$$t_i = \begin{cases} 1, & \text{if } a_i \in B; \\ 0, & \text{if } a_i \notin B. \end{cases}$$

Now if two subsets B and C are distinct, then there must be at least one element of A that is in one subset that is not in the other. Therefore at least one digit of $f(B)$ will differ from the corresponding digit in $f(C)$. Therefore f is one-to-one. In fact, our map is also onto. Every n-tuple $\sigma \in \{0, 1\}^n$ defines a subset B_σ with elements exactly those that correspond to the digits of σ that equal 1. So $f(B_\sigma) = \sigma$.

6.13.   The statement is true. *Extension:* For now we do not have the tools to extend this result, but once we can work with the cardinalities of infinite sets, this result will hold for infinite sets as well.

Proof of original statement: The answer to Challenge 6.12 gives a one-to-one, onto function from $\mathcal{P}(A)$ to $\{0, 1\}^n$; therefore, by the salvage to Challenge 6.10, we have that $|\mathcal{P}(A)| = |\{0, 1\}^n|$. Because $|\{0, 1\}| = 2$, there are two choices for each component in an n-tuple of $\{0, 1\}^n$. Therefore $|\{0, 1\}^n| = 2^n$, and the result follows.

6.14.   This result is very subtle, and the proof is difficult. Encourage your more ambitious students to consider the hint and give it a good effort, but don't let them be discouraged if they come up short. Students may find Challenge 6.15 more accessible, because it is a particular case of Challenge 6.14. Also note that the solution for Challenge 6.15 shows a specific example of the technique used to prove the general theorem. Thus, working through the solution to Challenge 6.15 first may make the proof for Challenge 6.14 easier to follow.

Proof: Observe that because g is one-to-one, there will be at most one value in any preimage $g^{-1}(a)$ for $a \in A$. Let $A' = A \setminus g(B)$. Now define $h:A \to B$ as follows:

$$h(a) = \begin{cases} f(a), & \text{if } a \in (g \circ f)^n(A') \text{ for some nonnegative integer } n; \\ g^{-1}(a), & \text{otherwise.} \end{cases}$$

First note that $(g \circ f)^n$ denotes $g \circ f$ composed with itself n times, where the case of $n = 0$ gives the identity function. Where did this function come from? According to the hint given with the challenge, for each $a \in A$, we will define $h(a)$ to be either $f(a)$ or $g^{-1}(a)$. If $a \notin g(B)$, then we must use f to define h. So for elements $a \notin g(B)$—that is, elements a in the set we defined above as A'—we must let $h(a) = f(a)$. (This corresponds to the case when $n = 0$ in the original definition of h above.) The remaining elements of A lie in $g(B)$. It would be great if we could just let $h(a) = g^{-1}(a)$ for these elements. But what if there were some $a \in (g \circ f)(A')$? Then $g^{-1}(a) \in f(A')$, and we might bump into values of h we already used for elements of A'. This would spoil the one-to-one property of h. Therefore, for any elements of A that lie in $(g \circ f)(A')$, we again let $h(a) = f(a)$. (This corresponds to the case above when $n = 1$). Similarly, if $a \in ((g \circ f) \circ (g \circ f))(A') = (g \circ f)^2(A')$, then $g^{-1}(a) \in f((g \circ f)(A'))$. If we use $g^{-1}(a)$ to define $h(a)$ in this case, we again might overlap with previously used values of h. So for $a \in (g \circ f)^2(A')$, we define $h(a) = f(a)$. (This corresponds to the case above when $n = 2$). We continue this process to yield the function h.

To finish the proof, we need to show that h is one-to-one and onto. To show h is one-to-one, we consider three cases. If $f(a_1) = f(a_2)$, then, because f is one-to-one, we have $a_1 = a_2$. Now suppose $g^{-1}(a_1) = g^{-1}(a_2)$. Because g is one-to-one by the salvage of Challenge 6.7, we know that g^{-1} is a function from $g(B)$ to B and that $(g \circ g^{-1})(y) = y$ for all $y \in g(B)$. Therefore $(g \circ g^{-1})(a_1) = (g \circ g^{-1})(a_2)$, and we have $a_1 = a_2$. The last case to consider is when $f(a_1) = g^{-1}(a_2)$. Note, however, that h was defined precisely so that g^{-1} was never used on any $a \in A$ for which f was used. So this last case never happens.

To show h is onto, consider $b_0 \in B$. We need to show that b_0 is the image of some element $a_0 \in A$ under h. First note that $g(b_0) \in A$ and, by our definition of A', $g(b_0) \notin A'$. If there is no nonnegative integer n for which $g(b_0) \in (g \circ f)^n(A')$, then we define $a_0 = g(b_0)$. Then by our definition of h, we have $h(a_0) = g^{-1}(a_0) = g^{-1}(g(b_0)) = b_0$, and we are done. Now we consider the case where $g(b_0) \in (g \circ f)^n(A')$ for some $n \geq 0$. Because we know that $g(b_0) \notin A'$, we actually have $n > 0$. So now we have $g(b_0) = (g \circ f)^n(a')$ for some $a' \in A'$ and $n > 0$. That is, we have $g(b_0) = g(f(g \circ f)^{n-1}(a'))$. But we know g is one-to-one; therefore $b_0 = f(g \circ f)^{n-1}(a')$. So b_0 is the image under h of $(g \circ f)^{n-1}(a')$, which is clearly an element of A because $n - 1 \geq 0$.

Thus, at last, we are able to conclude that h is one-to-one and onto. Whew! A serious effort, but this clever result is worth it.

6.15.   Let A denote $[0, 1]$ and let B denote $[0, 1)$. Following the hint in Appendix 1, define $f : A \to B$ by $f(x) = x/2$ and define $g : B \to A$ by $g(x) = x$. Each of these functions is clearly one-to-one. We will apply the method used to prove the Schroeder-Bernstein Theorem from Challenge 6.14 to construct a one-to-one, onto function $h : A \to B$. Using the notation from the proof of Challenge 6.14, $A' = A \setminus g(B) = \{1\}$. So we define $h(1) = f(1) = 1/2$. The remaining elements of A lie in $g(B)$, so we would like to let $h(a) = g^{-1}(a)$ for these elements, but unfortunately $g^{-1}(1/2) = 1/2$. This would give $h(1) = h(1/2) = 1/2$, which would spoil our plan to make h one-to-one. So we let $h(1/2) = f(1/2) = 1/4$. (Note what went wrong: $1/2 = (g \circ f(1))$. So when we apply g^{-1} to $1/2$, we end up in $f(A')$ and repeat a value already used by h.) But now we can't use g^{-1} to define $h(1/4)$, because $g^{-1}(1/4) = 1/4$, again spoiling the one-to-one property of h. So we let $h(1/4) = f(1/4) = 1/8$. (Note that $1/4 = (g \circ f \circ g \circ f)(1) = (g \circ f)^2(1)$. This is helpful in understanding the proof of the general theorem in Challenge 6.14.)

We continue this process, yielding a function h that maps $1/2^n$ to $1/2^{n+1}$ for all integers $n \geq 0$. Note, however, that for any x that does not equal one of these powers of $1/2$, we are safe in defining $h(x) = g^{-1}(x) = x$. Therefore h is defined as follows:

$$h(x) = \begin{cases} f(x) = x/2, & \text{if } x = 1/2^{n+1} \text{ for some integer } n \geq 0; \\ g^{-1}(x) = x, & \text{otherwise.} \end{cases}$$

The function h is one-to-one and onto by the arguments in the proof of Schroeder-Bernstein Theorem proved in Challenge 6.14. The schematic below illustrates the action of h.

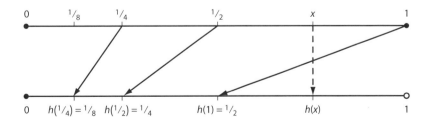

Stepping back

Under the given conditions, we would know $f(n)$ for all integers. Here's the most general claim:

Claim: Suppose the function $f: \mathbb{Z} \to \mathbb{R}$ satisfies $f(n + m) = f(n) + f(m)$ for all n and m in $\mathbb{Z}$. If $f(k)$ is known for a single integer k, then $f(n)$ is known for all integers.

Proof: First note that the given property of f can be applied sufficient times to obtain

$$f(k) = f(1 + 1 + \cdots + 1) = f(1) + f(1) + \cdots + f(1) = kf(1),$$

and thus $f(1) = (1/k)f(k)$. By the same argument, we have $f(n) = nf(1) = (n/k)f(k)$ for any positive integer n. Note also that $f(1) = f(1+0) = f(1) + f(0)$ and thus $f(0) = 0$. Finally we have $0 = f(0) = f(n + (-n)) = f(n) + f(-n)$. Therefore $f(-n) = -f(n)$, which gives function values for all negative integers, thus completing the proof.

7

Infinity
Understanding the unending

Infinity can be one of the most provocative topics in any mathematics course. There is much potential here for lively class discussion, if you choose to devote the class time. This module introduces students to the definition of an infinite set and proceeds with methods for comparing the "sizes" of infinite sets. Note that although finite sets have appeared earlier in the text (Module 6), this is the first time the phrase "finite set" is defined. This material works well at several levels, from an introduction to proofs or higher mathematics all the way to a capstone course for majors.

Make sure students see the importance of the definition of two sets having the same cardinality, but also reassure them that they don't always need to define a one-to-one, onto function to show $|A| = |B|$. In particular, to show a set A is *countable* (i.e., has the same cardinality as $\mathbb{N}$), they need only show that the elements of A can be listed ($a_1, a_2, a_3, \ldots$) in such a way that each element of A appears exactly once on the list. Note that such a list, in fact, defines a one-to-one, onto map $f: A \to \mathbb{N}$—namely, $f(a_n) = n$.

Special note: Challenge 7.12 actually poses the Continuum Hypothesis to students (see the "solution"). Decide how you want to handle this: You can tell students ahead of time that this question has been proven to have no answer, or you can let them struggle a bit before revealing the truth in what would certainly be a lively discussion. Emphasize that the status of the Continuum Hypothesis demonstrates that even within the rigorous bounds of mathematics, there are questions that cannot be answered. In defense of that very rigor, you can also point out that, at least in this case, mathematicians can *prove* the question is unanswerable.

The module ends with some fun paradoxes involving cleverly (or poorly!) defined sets and self-referential logic. The paradoxes reveal the need for constraints on how individual sets or logical statements are constructed in mathematics.

Solutions

7.1.   *Proof:* Following the hint in Appendix 1, we suppose $\mathbb{N}$ is finite. Thus there is set $S = \{1, 2, \dots, n\}$ for some positive integer n so that $|\mathbb{N}| = |S|$. Therefore we have a one-to-one, onto function $f: S \to \mathbb{N}$. Because $f(S)$ consists exactly of elements of the form $f(i)$, where $i = 1, 2, \dots, n$, we must have $|f(S)| \le n$. Thus, because $f(S) = \mathbb{N}$ has at most n elements, $\mathbb{N}$ must have a largest element, say, M. But $M + 1$ is also in $\mathbb{N}$, which contradicts f being onto. Therefore $\mathbb{N}$ is infinite.

Alternate proof: Here we use Cantor's definition of an infinite set. Let $\mathbb{E}$ be the set of even natural numbers; so $\mathbb{E}$ is a proper subset of $\mathbb{N}$. Consider $f: \mathbb{N} \to \mathbb{E}$, given by $f(n) = 2n$. Then f is one-to-one because $2n_1 = 2n_2$ implies $n_1 = n_2$, and f is onto because for any $m \in \mathbb{E}$, $m/2 \in \mathbb{N}$ and $f(m/2) = m$. Thus, because there is a one-to-one, onto function from $\mathbb{N}$ to a proper subset of itself, we have that $\mathbb{N}$ is infinite.

7.2.   The statement is a theorem. *Extension:* Let $k \ge 0$ be an integer. Then the set $S = \{-k, -(k-1), \dots, -2, -1, 0\} \cup \mathbb{N}$ has the same cardinality as $\mathbb{N}$.

Proof of extension: Define $f: S \to \mathbb{N}$ by $f(n) = n + k + 1$. The function f shifts all the elements of S exactly $k + 1$ units to the right. We see that f is one-to-one because $f(n_1) = f(n_2)$ implies $n_1 + k + 1 = n_2 + k + 1$, which implies $n_1 = n_2$. We also see that f is onto, because for any $m \in \mathbb{N}$, $m - k - 1 \in S$ and $f(m - k - 1) = m$. Because f is one-to-one and onto, we have $|S| = |\mathbb{N}|$.

Remark: Students may extend Challenge 7.2 in different ways. For example, they may show that the union of $\mathbb{N}$ with any finite set of other integers will have the same cardinality as $\mathbb{N}$. The extension given above is nice because an appropriate one-to-one, onto function is easy to define.

Students may be fascinated with the idea that there are ways to add elements to a set without changing its "size."

7.3.   The statement is a theorem. *Extension:* There is no immediately natural extension here, but one larger leap is Challenge 7.5.

Proof: We define $f: \mathbb{Z} \to \mathbb{N}$ as follows:

$$f(n) = \begin{cases} 2n, & \text{if } n > 0; \\ 1, & \text{if } n = 0; \\ 2(-n)+1, & \text{if } n < 0. \end{cases}$$

So f is a "shuffling map," sending 0 to 1, sending all the positive integers to even positive integers, and sending all the negative integers to odd positive integers greater than or equal to 3. Clearly f is one-to-one and onto, so $|\mathbb{Z}| = |\mathbb{N}|$.

7.4. The statement is a theorem. *Extension:* See Challenge 7.5.

Proof: Following the hint, we construct a list that contains each element of $\mathbb{Q}$ (the rational numbers) exactly once: q_1, q_2, q_3, … . Once we have such a list, the function $f:\mathbb{N} \to \mathbb{Q}$ defined by $f(n) = q_n$ is clearly one-to-one and onto.

There are many ways to construct a list of the rational numbers. We start naively by listing the natural numbers: 1, 2, 3, …. We will add the rest of the rationals in stages. Note that at any stage, we only add rationals that are not already on the list. First, after each natural number, we insert all fractions less than 1 with that number as a denominator:

$$1, 2, \mathbf{1/2}, 3, \mathbf{1/3}, \mathbf{2/3}, 4, \mathbf{1/4}, \mathbf{3/4}, 5, \mathbf{1/5}, \mathbf{2/5}, \mathbf{3/5}, \mathbf{4/5}, \ldots .$$

Note that for clarity, the newly inserted numbers are in boldface. To include the fractions greater than 1, after each number in the current list, we insert its reciprocal, provided it does not already appear in the list:

$$1, 2, 1/2, 3, 1/3, 2/3, \mathbf{3/2}, 4, 1/4, 3/4, \mathbf{4/3}, 5, 1/5, 2/5, \mathbf{5/2}, 3/5, \mathbf{5/3}, 4/5, \mathbf{5/4}, \ldots .$$

At this point, our list contains each positive rational number exactly once. To complete the list, we insert 0 at the beginning and all the negative rationals as shown:

$$\mathbf{0}, 1, \mathbf{-1}, 2, \mathbf{-2}, 1/2, \mathbf{-1/2}, 3, \mathbf{-3}, 1/3, \mathbf{-1/3}, 2/3, \mathbf{-2/3}, 3/2, \mathbf{-3/2}, 4, \mathbf{-4}, \ldots .$$

We use this list to define our one-to-one correspondence, and our proof is complete.

Remark: By this time, if not sooner, students may be reeling from the the implications of Cantor's definition of "equally numerous." There seem to be so many more rational than natural numbers; yet because we can construct an orderly method for listing the rationals, the two sets have the same cardinality. This is a good moment to pause, if class time is planned, and let students reflect on what it means for a set to be countable.

There are other ways to construct a list of the rationals. A popular method involves zig-zagging through a table of all possible fractions. Following the pattern of arrows in the figure, skipping repeats, we construct the following list of the rationals:

$$0, 1, -1, -1/2, -2, 2, 1/2, 1/3, 3, -3, -1/3, -1/4, -2/3, -3/2, -4, 4, \ldots .$$

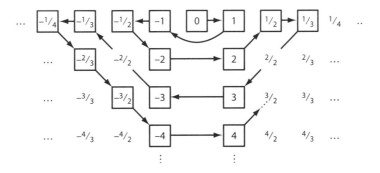

7.5. The statement is a theorem. *Extension:* If $n \geq 2$ is an integer and $S_1, S_2, \ldots, S_n$ are countable sets, then the set $S = S_1 \times S_2 \times \cdots \times S_n$ is countable.

Proof of extension: We will use mathematical induction. The base case is $n = 2$. Suppose S_1 and S_2 are countable. Then there is a one-to-one, onto function from the natural numbers to S_1, so we can write $S_1 = \{a_1, a_2, a_3, \ldots\}$. Similarly we can write $S_2 = \{b_1, b_2, b_3, \ldots\}$. To prove $S_1 \times S_2$ is countable, we need only describe a method of listing all the elements (a_i, b_j) so that each pair occurs exactly once. There are many ways to do this. Here is one, where the ordering is according to the sum of the indices of each pair:

$$S_1 \times S_2 = \{(a_1, b_1), (a_1, b_2), (a_2, b_1), (a_1, b_3), (a_2, b_2), (a_3, b_1), (a_1, b_4), (a_2, b_3), (a_3, b_2), (a_4, b_1), \ldots\}.$$

Note that we need only list enough elements in $S_1 \times S_2$ so that the pattern is clear.

We now assume that the result holds for k countable sets and consider $k + 1$ countable sets: $S_1, S_2, \ldots, S_{k+1}$, where $k \geq 1$. To show that $S = S_1 \times S_2 \times \cdots \times S_{k+1}$ is countable, note that we have $S' = S_1 \times S_2$ countable by our base case argument. Then $S = S' \times S_3 \times \cdots \times S_{k+1}$, which is a product of k countable sets. Therefore by our induction assumption, we have that S is countable. By mathematical induction, the result holds.

7.6. *Proof:* For convenience, let S denote the set $\{1, 2, \ldots, N\}$. We consider an arbitrary one-to-one function $f: S \to \mathcal{B}_N$. So for each $k \in S$, $f(k)$ is an n-tuple of 0's and 1's. We will show that f is not onto by finding an n-tuple of 0's and 1's that is not in $f(S)$. Let $\sigma = (s_1, s_2, \ldots, s_N)$, where

$$s_i = \begin{cases} 1, & \text{if the } i\text{th component of } f(i) = 0; \\ 0, & \text{if the } i\text{th component of } f(i) = 1. \end{cases}$$

Thus σ cannot be in $f(S)$, because it differs from each n-tuple in $f(S)$ in at least one component. Therefore f is not onto.

7.7. $\mathcal{B}_\infty$ is not countable. By proving this statement, we are proving that there are different sizes of infinity.

Proof: Suppose $\mathcal{B}_\infty$ is countably infinite. Then we can write $\mathcal{B}_\infty = \{\sigma_1, \sigma_2, \sigma_3, \ldots\}$, where each σ_i is an unending vector of 0's and 1's. We will now construct an element τ of $\mathcal{B}_\infty$ that is not on the list. Let the components of $\tau = (t_1, t_2, \ldots)$ be defined as follows:

$$t_i = \begin{cases} 1, & \text{if the } i\text{th component of } \sigma_i = 0; \\ 0, & \text{if the } i\text{th component of } \sigma_i = 1. \end{cases}$$

We note that for each index i, τ differs from σ_i in the ith component. Thus τ does not equal any σ_i on the list, and therefore it does not appear on the list. But by our original assumption, our list contained all elements of our set. This contradiction implies no such list can exist. Thus $\mathcal{B}_\infty$ is not countable.

Remark: Be alert to those students who may want to "fix" the list by adding the missing vector, say at the beginning of the list. If this suggestion is made, invite other students to respond. Some may say that adding the missing vector to the list doesn't help, because the argument could just be used again to create another vector that is not on the new list. Although this is a valid response, note that it misses the real point of the contradiction. The list was assumed to be complete at the beginning, so by finding a single missing vector, we have shattered the assumption that the set is countable.

The method of proof here is called Cantor's diagonalization argument. Cantor's work was highly controversial. Many mathematicians resisted his arguments about infinite sets; some even attacked him in personal ways. As a result of the stress, along with some other psychological issues, Cantor spent many of his later years in a sanatorium.

7.8.   The answer is "No." $\mathbb{R}$ is not countable.

Proof: We will show that there is no way to list the real numbers between 0 and 1 and, therefore, no way to list all the reals. Let S denote the set of real numbers x, $0 \le x < 1$. Then each $x \in S$ can be written as $x = 0.d_1d_2d_3 \ldots$, where each d_i is a digit from 0 to 9. As in Challenge 7.7, we assume the set S is countable, so we have a list $s_1, s_2, s_3, \ldots$ in which each element of S appears exactly once:

$$s_1 = 0.d_{1,1}d_{1,2}d_{1,3} \cdots$$
$$s_2 = 0.d_{2,1}d_{2,2}d_{2,3} \cdots$$
$$s_3 = 0.d_{3,1}d_{3,2}d_{3,3} \cdots$$
$$\vdots$$

We apply Cantor's diagonalization argument to construct a number in S that does not appear on the list. Define the digits of $t = 0.t_1t_2t_3 \ldots$ as follows:

$$t_i = \begin{cases} 1, & \text{if } d_{i,i} = 0; \\ 0, & \text{if } d_{i,i} \ne 0. \end{cases}$$

So for each index i, t differs from s_i in the ith digit, and therefore t cannot equal any of the numbers in the list. But by our original assumption, our list contained all elements of S. This contradiction implies no such list can exist and so the set S is uncountable.

Remark: Note that in our construction of t, we had many choices for how to distinguish t from each s_i. We could have chosen any two digits to define t—changing 4's to 7's and 7's to 4's, for example.

7.9.   The statement is a theorem. *Extension:* Let S be a set. Then $|\mathcal{P}(S)| = 2^{|S|}$. (See *Remark* below.)

Proof of original statement: We will present a general proof that works for any set S, but note that we already have this result for finite S. By Challenge 6.13, we know that $|\mathcal{P}(S)| = 2^{|S|}$, which is larger than $|S|$. (This is easy to prove by induction.)

Now suppose S is infinite. What does it mean to show that $|\mathcal{P}(S)| > |S|$? Suppose that $f: S \to \mathcal{P}(S)$. We will show that f can never be onto. In particular, we will show that there are elements of $\mathcal{P}(S)$—that is, subsets of S—that are not equal to the image of any $s \in S$ under f. First note that the function f maps each $s \in S$ to a subset $f(s)$ of S. Now consider the subset $T \subseteq S$ defined as $T = \{s : s \notin f(s)\}$. So, for each $s \in S$, if s lies in its corresponding subset $f(s)$, then s will not be in T. If s does not lie in its corresponding subset, then s will be in T. Now we will show that no element of S is mapped to T, and thus f is not onto.

Suppose there is some $t \in S$ for which $f(t) = T$. The question is, does t lie in T? According to the definition of T, t will be in T if and only if t is not in T, which is clearly a contradiction. Thus T is not in the image of f; therefore f is not onto. Thus $\mathcal{P}(S)$ is bigger than S. Note that because we never needed to know whether S was finite or infinite, this proof works in general.

Remark: As stated in the proof, for finite S, the extension is simply a restatement of Challenge 6.13. But what does the extension mean if S is infinite? What does it mean to raise 2 to an infinite power? We are treading into the land of transfinite arithmetic, which is beyond the scope of this text. Suffice it to say that mathematicians define the value of $2^{|S|}$ to be the cardinality of $\mathcal{P}(S)$. Thus our extension is valid for infinite sets by definition.

7.10.   There is a set with cardinality larger than $\mathbb{R}$—namely, $\mathcal{P}(\mathbb{R})$, the power set of $\mathbb{R}$. A set with larger cardinality than this is $\mathcal{P}(\mathcal{P}(\mathbb{R}))$, the power set of the power set of $\mathbb{R}$. Elements of $\mathcal{P}(\mathcal{P}(\mathbb{R}))$ are subsets consisting of subsets of $\mathbb{R}$, such as $\{\{1, 2, 3\}, \{\sqrt{2}\}, \mathbb{Q}\}$.

7.11.   There does not exist an infinite set with cardinality larger than all other infinities. To anyone who claims that there is such a set, ask that person to consider its power set and apply Challenge 7.9.

7.12.   Amazingly enough, it has been proven that this question is not answerable using the standard axioms of mathematics. At one time, mathematicians conjectured that there is no cardinality between $|\mathbb{N}|$ and $|\mathbb{R}|$. This conjecture is known as the *Continuum Hypothesis,* because the set of real numbers is sometimes referred to as the *continuum.* In 1940, Kurt Gödel proved that it is impossible to disprove the Continuum Hypothesis using the standard axioms and logical techniques of mathematics. In 1963, Paul Cohen proved that it is likewise impossible to prove the Continuum Hypothesis. Thus the truth or falseness of this famous conjecture lies outside the bounds of our entire mathematical structure. Thus no matter what the students guessed, they are correct!

7.13.   *Proof:* We show that the number of algebraic numbers is countable and thus cannot contain all of $\mathbb{R}$, which is uncountable. Let $\mathbb{Z}[x]$ denote the set of all polynomials with integer coefficients. We will first show that $\mathbb{Z}[x]$ is countable by describing a way to list all the elements in $\mathbb{Z}[x]$. At each stage of the process, we will be careful not to repeat any polynomials that were listed at an earlier stage.

We put the zero polynomial first on the list. Then we list all polynomials of degree 1 or less with coefficients no larger than 1 in absolute value. Note that there are only finitely many of these. Next we list all the polynomials of degree 2 or less with coefficients no larger than 2 in absolute value.

(Again there are only a finite number.) Then we list all the polynomials of degree 3 or less with coefficients no larger than 3 in absolute value, and so forth. Because there are only finitely many polynomials to add to the list at each stage, because every polynomial has finite degree, and because every polynomial has a maximum coefficient in absolute value, we know that every polynomial must eventually appear on the list at some point.

Now that we know $\mathbb{Z}[x]$ is countable, we will use the fact that each polynomial in $\mathbb{Z}[x]$ has only finitely many zeros (at most n, where n is the degree of the polynomial). We can now construct a list of all algebraic numbers as follows: List the zeros of the first polynomial on our list of $\mathbb{Z}[x]$, then list the zeros of the second polynomial, then the zeros of the third, and so on, skipping any repeated zeros. Because each polynomial has only a finite number of zeros, we know that each algebraic number must eventually appear on the list. Thus the set of algebraic numbers is a countable subset of $\mathbb{R}$, which is uncountable. Therefore there do exist transcendental numbers. (In fact, there must be an uncountable number of them.)

7.14.   To show that T has infinitely many elements, we will show that $S = \mathbb{N} \setminus T$ is finite. Recall that S is the set of all natural numbers that are describable in English words using no more than 50 characters, where we are allowed to use only the 26 letters of the alphabet and the space character. This gives us 27 possible characters, so we have at most 27 choices for each of 50 positions in our description, for a total of at most 27^{50} descriptions. Many of these will not even make sense in English, so we have that S has many fewer than 27^{50} elements. Therefore T is infinite.

To show that T has a smallest element, note that $T \subseteq \mathbb{N}$; so all elements of T are at least 1. If we put the elements of T in increasing order, then the first element on the list will be the smallest element.

The number t can be described as "the smallest number in T," which requires only 24 characters. Therefore $t \in S$. But by the definition of T, we also have $t \notin S$. This contradiction suggests that the set S cannot be defined in a meaningful way.

Remark: The proof that T has a smallest element could be more rigorous. Technically we should use the *Well-Ordering Principle,* which states that any nonempty subset of the natural numbers has a smallest element. This principle is easy to prove using induction.

7.15.   Let's try to answer the question "Does the barber shave himself?" The barber shaves a man if and only if that man does not shave himself. Therefore the barber shaves himself if and only if the barber does not shave himself. Thus the question has no meaningful answer and is a paradox. Moreover, there cannot exist such a village with such a barber. (If students suggest that perhaps the barber does not shave at all and has a heavy beard, see if another student can point out that the conditions in this village make it clear that all men are shaved, either by themselves or by the barber.)

Looking at the set *NoWay* of all sets that do not contain themselves as elements, we see that a set $S \in NoWay$ if and only if $S \notin S$. But then $NoWay \in NoWay$ if and only if $NoWay \notin NoWay$, which is self-contradictory and, hence, another paradox.

Russell's paradox reveals the need for some constraints on how sets are defined in mathematics. In particular, we do not allow sets that are elements of themselves.

Stepping back

The statement cannot be false, for if it is, then, in particular, the statement "This statement is false" is false. This implies the original statement is true, and we have a contradiction. Thus the original statement must be true.

For an "or" statement to be true, at least one of the claims must be true. Because we have shown that the first claim is false, the second must be true. So "I will get an A in my math class" is true.

We can modify the sentence to prove any claim P, even if P is completely ludicrous. Applying the reasoning used above, the statement "This sentence is false or P" is true, which implies the statement P is true. The moral of the story is that self-referential logic can lead to problems. If you don't believe that self-referential logic is a problem, then return to the beginning of this sentence and read it again.

8

Recursively defined functions
The next generation

Recursively defined functions are very important in discrete mathematics and computer science. After beginning with some simple examples, this module introduces the idea of a generating function as a general method for finding closed-form solutions to recurrence relations. Generating functions are formal power series with, in the context of this module, geometric series playing a central role. The technique required for solving recurrence relations offers excellent practice for formal manipulation of power series. Though this topic is most often associated with discrete math, this module's generating function approach also develops skills that are important in other areas of mathematics. The module ends by having students discover the *characteristic polynomial* of a recurrence relation and an alternative approach to finding closed-form solutions. This approach is computationally simpler than using generating functions, but establishing this technique requires the use of generating functions.

Solutions

8.1. We will use ad hoc methods to find closed formulas for these sequences. Mainly, we iterate the recurrence formula to generate terms until we see a pattern.

- $a_n = na_{n-1} = n(n-1)a_{n-2} = n(n-1)(n-2)a_{n-3} = \cdots = n(n-1)(n-2) \cdots 3 \cdot 2 \cdot 1 \cdot 1 = n!$
- $b_n = b_{n-1} + n = b_{n-2} + (n-1) + n = b_{n-3} + (n-2) + (n-1) + n = \cdots = \sum_{i=1}^{n} i = n(n+1)/2$
- $c_n = 3c_{n-1} = 3^2 c_{n-2} = \cdots = 3^n c_0 = 3^n$
- $d_n = \frac{1}{2}(d_{n-1} + d_{n-2})$, with $d_0 = d_1 = 1$, gives us $d_2 = 1$; so clearly $d_n = 1$ in general.

Remark: The closed formulas for the first two sequences above may challenge some students. For the second sequence, the formula $b_n = n(n+1)/2$ requires that they know how to find the sum of the first n natural numbers. More subtly, the formula for the first sequence, $a_n = n!$, actually has "implied" dots, because $n!$ is just mathematical shorthand for $n(n-1)(n-2) \cdots 3 \cdot 2 \cdot 1$. You might acknowledge this to your students with a sense of irony or a wink of amusement.

8.2.   We are given that $a_1 = 1$ and for all $n \geq 2$, $a_n = 4a_{n-1} - 2$. Using the recursive relation, we find the first five terms of the sequence are

$$a_1 = 1$$
$$a_2 = 4(a_1) - 2 = 4 \cdot 1 - 2$$
$$a_3 = 4(a_2) - 2 = 4(4 \cdot 1 - 2) - 2 = 4^2 - 4 \cdot 2 - 2$$
$$a_4 = 4(a_3) - 2 = 4(4^2 - 4 \cdot 2 - 2) - 2 = 4^3 - 4^2 \cdot 2 - 4 \cdot 2 - 2$$
$$a_5 = 4(a_4) - 2 = 4(4^3 - 4^2 \cdot 2 - 4 \cdot 2 - 2) - 2 = 4^4 - 4^3 \cdot 2 - 4^2 \cdot 2 - 4 \cdot 2 - 2.$$

Note the benefit of leaving the values unsimplified. We can now guess that a_n satisfies

$$a_n = 4^{n-1} - 4^{n-2} \cdot 2 - \cdots - 4^2 \cdot 2 - 4 \cdot 2 - 2 = 4^{n-1} - 2(4^{n-2} + \cdots + 4^2 + 4 + 1).$$

Because a^n appears to involve a finite geometric series, we recall that for $x \neq 1$,

$$1 + x + x^2 + \cdots + x^k = \frac{x^{k+1} - 1}{x - 1}.$$

(This formula follows immediately if we expand $(x - 1)(1 + x + x^2 + \cdots + x^k)$). Therefore we have

$$4^{n-2} + \cdots + 4^2 + 4 + 1 = \frac{4^{n-1} - 1}{4 - 1} = (1/3)(4^{n-1} - 1).$$

Thus our conjectured formula for a_n becomes

$$a_n = 4^{n-1} - 2(4^{n-2} + \cdots + 4^2 + 4 + 1) = 4^{n-1} - \frac{2}{3}(4^{n-1} - 1) = \frac{1}{3}(4^{n-1} + 2).$$

We now prove this formula is valid by using induction.

Proof: For the base case, we note that for $n = 1$, $(1/3)(4^{n-1} + 2) = 1$, which is the given value of a_1. Assuming the formula holds for $n = k$, we now consider a_{k+1}:

$$a_{k+1} = 4(a_k) - 2 \qquad \text{(recursive definition of } a_n\text{)}$$
$$= 4 \cdot \frac{1}{3}(4^{k-1} + 2) - 2 \qquad \text{(induction assumption)}$$
$$= \frac{1}{3} \cdot 4^k + \frac{2}{3} = \frac{1}{3}(4^k + 2).$$

Thus our formula holds for all $n \geq 1$.

Remark: Guessing a formula for a_n is nontrivial in this challenge. Students may need some hints, as well as a reminder about the formula for the sum of a finite geometric series.

8.3.   We know the sum of the geometric series is

$$\frac{1}{1 - x} = 1 + x + x^2 + x^3 + \cdots = \sum_{n=0}^{\infty} x^n.$$

Replacing x with ax, we obtain a power series for the first function:

$$\frac{1}{1-ax} = 1 + ax + \left(ax\right)^2 + \left(ax\right)^3 + \cdots = \sum_{n=0}^{\infty} a^n x^n.$$

We multiply the original geometric series by x to obtain the second function:

$$\frac{x}{1-x} = x\left(1 + x + x^2 + x^3 + \cdots\right) = x\sum_{n=0}^{\infty} x^n = \sum_{n=0}^{\infty} x^{n+1}.$$

For the last function, we write out the product as suggested in the hint (the result can be proven correct by induction):

$$\frac{1}{\left(1-x\right)^2} = \frac{1}{\left(1-x\right)}\frac{1}{\left(1-x\right)} = \left(1 + x + x^2 + x^3 + \cdots\right)\left(1 + x + x^2 + x^3 + \cdots\right)$$

$$= 1 + 2x + 3x^2 + 4x^3 + \cdots = \sum_{n=0}^{\infty}(n+1)x^n.$$

Remark: Some students may be bold or experienced enough to think of differentiating one power series to obtain another. Differentiation is another legitimate way to manipulate formal power series and offers an alternative derivation of the power series for the last function above. The function $1/(1-x)^2$ is the derivative of $1/(1-x)$, so we have

$$\frac{1}{\left(1-x\right)^2} = \frac{d}{dx}\left(\frac{1}{1-x}\right) = \frac{d}{dx}\left(1 + x + x^2 + x^3 + \cdots\right) = 1 + 2x + 3x^2 + 4x^3 + \cdots = \sum_{n=0}^{\infty}(n+1)x^n = \sum_{n=1}^{\infty} nx^{n-1}.$$

8.4. We factor $1/(4x^2 - 4x + 1)$ to obtain $1/(2x-1)^2 = 1/(1-2x)^2$. Then we replace x with $2x$ in the last series from Challenge 8.3 to obtain

$$\frac{1}{(1-2x)^2} = \sum_{n=0}^{\infty}(n+1)(2x)^n = 1 + 2(2x) + 3(2x)^2 + 4(2x)^3 + \cdots$$

$$= 1 + 4x + 12x^2 + 32x^3 + \cdots.$$

8.5. *Proof:* We are given that $f(x) = \sum_{n=0}^{\infty} a_n x^n$. We also notice that $5xf(x) = 5x\sum_{n=0}^{\infty} a_n x^n = \sum_{n=0}^{\infty} 5a_n x^{n+1} = \sum_{n=1}^{\infty} 5a_{n-1} x^n$. So we have

$$f(x) - 5xf(x) = \sum_{n=0}^{\infty} a_n x^n - \sum_{n=1}^{\infty} 5a_{n-1} x^n$$

$$= a_0 + \sum_{n=1}^{\infty} \left(a_n x^n - 5a_{n-1} x^n \right)$$

$$= a_0 + \sum_{n=1}^{\infty} \left(a_n - 5a_{n-1} \right) x^n$$

$$= a_0$$

because $a_n - 5a_{n-1} = 0$.

Thus we have $f(x) - 5xf(x) = f(x)(1 - 5x) = a_0 = 1$; so $f(x) = 1/(1 - 5x)$. This gives us a new formal power series $f(x) = \sum_{n=0}^{\infty} (5x)^n = \sum_{n=0}^{\infty} 5^n x^n$. So we have $\sum_{n=0}^{\infty} a_n x^n = \sum_{n=0}^{\infty} 5^n x^n$. Therefore $a_n = 5^n$ for $n = 0, 1, 2, \dots$, and we have our closed formula for a_n.

Remark: Students may notice that the closed formula for a_n in this challenge is very easily conjectured just by iterating the recurrence formula, as we did for the sequences in Challenge 8.1. Reassure them that the extra effort required by formal power series will allow them to solve more difficult sequences, where iteration alone would not reveal a pattern.

8.6. Let $f(x)$ be a generating function for the sequence $a_0, a_1, \dots$. We can rewrite the recurrence relation as $a_n - 2a_{n-1} + a_{n-2} = 0$, so we will follow the technique introduced in Challenge 8.5 and simplify the formal power series $f(x) - 2xf(x) + x^2 f(x)$. First observe that

$$f(x) = \sum_{n=0}^{\infty} a_n x^n = a_0 + a_1 x + \sum_{n=2}^{\infty} a_n x^n$$

$$2xf(x) = 2x\sum_{n=0}^{\infty} a_n x^n = \sum_{n=0}^{\infty} 2a_n x^{n+1} = \sum_{n=1}^{\infty} 2a_{n-1} x^n = 2a_0 x + \sum_{n=2}^{\infty} 2a_{n-1} x^n$$

$$x^2 f(x) = x^2 \sum_{n=0}^{\infty} a_n x^n = \sum_{n=0}^{\infty} a_n x^{n+2} = \sum_{n=2}^{\infty} a_{n-2} x^n .$$

Therefore we have

$$f(x) - 2xf(x) + x^2 f(x) = a_0 + a_1 x + \sum_{n=2}^{\infty} a_n x^n - 2a_0 x - \sum_{n=2}^{\infty} 2a_{n-1} x^n + \sum_{n=2}^{\infty} a_{n-2} x^n$$

$$= a_0 + (a_1 - 2a_0)x + \sum_{n=2}^{\infty} (a_n - 2a_{n-1} + a_{n-2})x^n$$

$$= a_0 + (a_1 - 2a_0)x \qquad\qquad \text{(because } a_n - 2a_{n-1} + a_{n-2} = 0\text{)}$$

$$= 3 - 8x .$$

Now we have $f(x)(1 - 2x + x^2) = 3 - 8x$, so $f(x) = \frac{3-8x}{1-2x+x^2} = \frac{3-8x}{(1-x)^2}$. Using the last power series derived in Challenge 8.3, we have

$$f(x) = (3 - 8x)\frac{1}{(1-x)^2} = (3 - 8x)\sum_{n=0}^{\infty}(n+1)x^n$$

$$= \sum_{n=0}^{\infty}3(n+1)x^n + \sum_{n=0}^{\infty}(-8)(n+1)x^{n+1}$$

$$= 3 + \sum_{n=1}^{\infty}3(n+1)x^n + \sum_{n=1}^{\infty}(-8)nx^n = 3 + \sum_{n=1}^{\infty}(3-5n)x^n$$

$$= \sum_{n=0}^{\infty}(3-5n)x^n .$$

Finally, because $f(x)$ was defined to be $\sum_{n=0}^{\infty}a_n x^n$, we have $a_n = 3 - 5n$ for all $n = 0, 1, 2, \ldots$. Thus we have a closed-form expression for a_n. Encourage students to check their formula for early values of a_n.

8.7. We have $f(x)$, a generating function for the sequence $a_0, a_1, \ldots$, with $a_0 = 1$, $a_1 = 4$, and $a_n = 4a_{n-1} - 4a_{n-2}$ for all $n \geq 2$. We rewrite the recurrence relation as $a_n - 4a_{n-1} + 4a_{n-2} = 0$, so we will simplify the formal power series $f(x) - 4xf(x) + 4x^2f(x)$. Observe that

$$f(x) = \sum_{n=0}^{\infty}a_n x^n = a_0 + a_1 x + \sum_{n=2}^{\infty}a_n x^n = 1 + 4x + \sum_{n=2}^{\infty}a_n x^n$$

$$4xf(x) = 4x\sum_{n=0}^{\infty}a_n x^n = \sum_{n=0}^{\infty}4a_n x^{n+1} = \sum_{n=1}^{\infty}4a_{n-1}x^n = 4a_0 x + \sum_{n=2}^{\infty}4a_{n-1}x^n = 4x + \sum_{n=2}^{\infty}4a_{n-1}$$

$$4x^2 f(x) = 4x^2\sum_{n=0}^{\infty}a_n x^n = \sum_{n=0}^{\infty}4a_n x^{n+2} = \sum_{n=2}^{\infty}4a_{n-2}x^n .$$

Therefore we have

$$f(x) - 4xf(x) + 4x^2 f(x) = 1 + 4x + \sum_{n=2}^{\infty}a_n x^n - 4x - \sum_{n=2}^{\infty}a_{n-1}x^n + \sum_{n=2}^{\infty}4a_{n-2}x^n$$

$$= 1 + \sum_{n=2}^{\infty}(a_n - 4a_{n-1} + 4a_{n-2})x^n$$

$$= 1$$

$$\text{because } a_n - 4a_{n-1} + 4a_{n-2} = 0 .$$

Now we have $f(x)(1 - 4x + 4x^2) = 1$, so $f(x) = \frac{1}{1-4x+4x^2} = \frac{1}{(1-2x)^2}$. Using the power series derived in Challenge 8.4, we have

$$f(x) = \frac{1}{(1-2x)^2} = \sum_{n=0}^{\infty}(n+1)(2x)^n = \sum_{n=0}^{\infty}(n+1)2^n x^n .$$

Because $f(x) = \sum_{n=0}^{\infty}a_n$, we have $a_n = (n+1)2^n$.

8.8.   For convenience, we will shift our sequence to begin with a_0 rather than a_1. Thus we consider $f(x)$, a generating function for the sequence $a_0, a_1, \ldots$, with $a_0 = 1$ and $a_n = 4a_{n-1} - 2$ for all $n \geq 1$. This time we rewrite the recurrence relation as $a_n - 4a_{n-1} = -2$, and we simplify the formal power series $f(x) - 4xf(x)$. Observe that

$$f(x) = \sum_{n=0}^{\infty} a_n x^n = a_0 + \sum_{n=1}^{\infty} a_n x^n = 1 + \sum_{n=1}^{\infty} a_n x^n$$

$$4xf(x) = \sum_{n=0}^{\infty} 4a_n x^{n+1} = \sum_{n=1}^{\infty} 4a_{n-1} x^n .$$

Therefore we have

$$f(x) - 4xf(x) = 1 + \sum_{n=1}^{\infty} a_n x^n + \sum_{n=1}^{\infty} 4a_{n-1} x^n$$

$$= 1 + \sum_{n=1}^{\infty} (a_n - 4a_{n-1}) x^n = 1 + \sum_{n=1}^{\infty} -2x^n$$

because $a_n - 4a_{n-1} = -2$.

Using the formula for the sum of a geometric series, we also know that $\sum_{n=1}^{\infty} -2x^n = -2x \sum_{n=0}^{\infty} x^n = \frac{-2x}{1-x}$. So now we have

$$f(x) - 4xf(x) = (1 - 4x)f(x) = 1 + \frac{-2x}{1-x} = \frac{1-3x}{1-x},$$

which implies $f(x) = \frac{1-3x}{(1-4x)(1-x)}$.

We apply the method of partial fractions, as suggested in the hint. We want to find constants A and B so that

$$\frac{1-3x}{(1-4x)(1-x)} = \frac{A}{1-4x} + \frac{B}{1-x}.$$

Thus we want

$$\frac{A}{1-4x} + \frac{B}{1-x} = \frac{A(1-x) + B(1-4x)}{(1-4x)(1-x)} = \frac{1-3x}{(1-4x)(1-x)}.$$

So we must have equal numerators: $A(1-x) + B(1-4x) = 1 - 3x$. Letting $x = 1$ in this equation, we have $-3B = -2$; so $B = 2/3$. Similarly, letting $x = 1/4$, we have $(3/4)A = 1/4$; so $A = 1/3$. Therefore we have

$$f(x) = \sum_{n=0}^{\infty} a_n x^n = \frac{1/3}{1-4x} + \frac{2/3}{1-x} = \frac{1}{3}\sum_{n=0}^{\infty}(4x)^n + \frac{2}{3}\sum_{n=0}^{\infty}x^n = \sum_{n=0}^{\infty}\left(\frac{1}{3}4^n + \frac{2}{3}\right)x^n.$$

This implies $a_n = \frac{1}{3}4^n + \frac{2}{3} = \frac{1}{3}(4^n + 2)$, a formula equivalent to the one we discovered in Challenge 8.2, where the sequence started with $a_1 = 1$.

8.9.   We have $f(x)$, a generating function for the sequence F_0, F_1, … , with $F_0 = F_1 = 1$ and $F_n = F_{n-1} + F_{n-2}$ for all $n \geq 2$. We rewrite the recurrence relation as $F_n - F_{n-1} - F_{n-2} = 0$, and we simplify the formal power series $f(x) - xf(x) - x^2 f(x)$. As with similar challenges, we find that

$$f(x) = \sum_{n=0}^{\infty} F_n x^n = F_0 + F_1 x + \sum_{n=2}^{\infty} F_n x^n = 1 + x + \sum_{n=2}^{\infty} F_n x^n$$

$$xf(x) = x\sum_{n=0}^{\infty} F_n x^n = \sum_{n=0}^{\infty} F_n x^{n+1} = \sum_{n=1}^{\infty} F_{n-1} x^n = F_0 x + \sum_{n=2}^{\infty} F_{n-1} x^n = x + \sum_{n=2}^{\infty} F_{n-1} x^n$$

$$x^2 f(x) = x^2 \sum_{n=0}^{\infty} F_n x^n = \sum_{n=0}^{\infty} F_n x^{n+2} = \sum_{n=2}^{\infty} F_{n-2} x^n .$$

Therefore we have

$$f(x) - xf(x) - x^2 f(x) = 1 + x + \sum_{n=2}^{\infty} F_n x^n - x - \sum_{n=2}^{\infty} F_{n-1} x^n - \sum_{n=2}^{\infty} F_{n-2} x^n$$

$$= 1 + \sum_{n=2}^{\infty} (F_n - F_{n-1} - F_{n-2}) x^n = 1$$

because $F_n - F_{n-1} - F_{n-2} = 0$.

  Thus $f(x) = \frac{1}{1-x-x^2}$. To express $f(x)$ in terms of geometric series, we need to factor the denominator of this expression. Unfortunately the roots of the quadratic are irrational, so we use the quadratic formula to obtain $x = \frac{1 \pm \sqrt{1-4(-1)(1)}}{2(-1)} = \frac{1 \pm \sqrt{5}}{-2}$. We let $\alpha = \frac{1+\sqrt{5}}{-2}$ and $\beta = \frac{1-\sqrt{5}}{-2}$ and use the method of partial fractions to obtain

$$f(x) = \frac{1}{1-x-x^2} = \frac{-1}{x^2 + x - 1} = \frac{-1}{(x-\alpha)(x-\beta)} = \frac{A}{x-\alpha} + \frac{B}{x-\beta} = \frac{A(x-\beta) + B(x-\alpha)}{(x-\alpha)(x-\beta)} .$$

Setting numerators equal to obtain $-1 = A(x-\beta) + B(x-\alpha)$, we let $x = \alpha$, which yields $A = \frac{-1}{\alpha - \beta}$. We let $x = \beta$ to obtain $B = \frac{-1}{\beta - \alpha}$. $A = \frac{-1}{\alpha - \beta}$ simplifies to $A = 1/\sqrt{5}$, so $B = -1/\sqrt{5}$. So now we have

$$f(x) = \frac{1}{\sqrt{5}} \cdot \frac{1}{x-\alpha} + \frac{-1}{\sqrt{5}} \cdot \frac{1}{x-\beta} = \frac{-1}{\sqrt{5}} \cdot \frac{1}{\alpha - x} + \frac{1}{\sqrt{5}} \cdot \frac{1}{\beta - x} = \frac{-1}{\alpha\sqrt{5}} \cdot \frac{1}{1-x/\alpha} + \frac{1}{\beta\sqrt{5}} \cdot \frac{1}{1-x/\beta} .$$

Expanding this expression into geometric series, we have

$$f(x) = \frac{-1}{\alpha\sqrt{5}} \sum_{n=0}^{\infty} (x/\alpha)^n + \frac{1}{\beta\sqrt{5}} \sum_{n=0}^{\infty} (x/\beta)^n = \sum_{n=0}^{\infty} \frac{-1}{\alpha\sqrt{5}} (1/\alpha)^n x^n + \sum_{n=0}^{\infty} \frac{1}{\beta\sqrt{5}} (1/\beta)^n x^n$$

$$= \sum_{n=0}^{\infty} \left(\frac{-1}{\alpha\sqrt{5}} (1/\alpha)^n + \frac{1}{\beta\sqrt{5}} (1/\beta)^n \right) x^n = \sum_{n=0}^{\infty} \frac{1}{\sqrt{5}} \left((1/\beta)^{n+1} - (1/\alpha)^{n+1} \right) x^n .$$

  Whew! At last, we have a formula for F_n, the nth Fibonacci number:

$$F_n = \frac{1}{\sqrt{5}} \left((1/\beta)^{n+1} - (1/\alpha)^{n+1} \right),$$

where $\alpha = \frac{1+\sqrt{5}}{-2}$, $\beta = \frac{1-\sqrt{5}}{-2}$.

We can rewrite this expression by rationalizing denominators as follows.

$$\frac{1}{\alpha} = \frac{-2}{1+\sqrt{5}} = \frac{-2}{1+\sqrt{5}} \cdot \frac{1-\sqrt{5}}{1-\sqrt{5}} = \frac{-2}{-4} \cdot (1-\sqrt{5}) = \frac{1-\sqrt{5}}{2} \text{ and}$$

$$\frac{1}{\beta} = \frac{-2}{1-\sqrt{5}} = \frac{-2}{1-\sqrt{5}} \cdot \frac{1+\sqrt{5}}{1+\sqrt{5}} = \frac{-2}{-4} \cdot (1+\sqrt{5}) = \frac{1+\sqrt{5}}{2}.$$

Thus

$$F_n = \frac{1}{\sqrt{5}} \left(\left(\frac{1+\sqrt{5}}{2} \right)^{n+1} - \left(\frac{1-\sqrt{5}}{2} \right)^{n+1} \right)$$

for $n = 0, 1, 2, \ldots$.

Remark: We note two things about this last formula for F_n. First, students may find it interesting to verify that this expression, as well as the previous formula for F_n, actually has integer values for all $n = 0, 1, 2, \ldots$. Second, and more profoundly, the number $\frac{1}{\beta} = \frac{1+\sqrt{5}}{2}$ may look familiar. It is the golden ratio, usually denoted ϕ. This number is related to the Fibonacci numbers in the following way: If we take the limit of ratios of consecutive Fibonacci numbers, we get ϕ. That is

$$\lim_{n \to \infty} \frac{F_{n+1}}{F_n} = \phi.$$

We can prove this result by expanding the ratios $\frac{F_{n+1}}{F_n}$, using continued fractions, as follows:

$$\frac{F_1}{F_0} = \frac{1}{1} = 1$$

$$\frac{F_2}{F_1} = \frac{F_1 + F_0}{F_1} = 1 + \frac{F_0}{F_1} = 1 + \frac{1}{\frac{F_1}{F_0}} = 1 + \frac{1}{1}$$

$$\frac{F_3}{F_2} = \frac{F_2 + F_1}{F_2} = 1 + \frac{F_1}{F_2} = 1 + \frac{1}{\frac{F_2}{F_1}} = 1 + \frac{1}{1 + \frac{1}{1}}$$

$$\frac{F_4}{F_3} = \frac{F_3 + F_2}{F_3} = 1 + \frac{F_2}{F_3} = 1 + \frac{1}{\frac{F_3}{F_2}} = 1 + \frac{1}{1 + \frac{1}{1 + \frac{1}{1}}}$$

$$\vdots$$

If we assume for now that our limit exists, we have

$$\lim_{n \to \infty} \frac{F_{n+1}}{F_n} = \phi = 1 + \cfrac{1}{1 + \cfrac{1}{1 + \cfrac{1}{\ddots}}} \quad,$$

which implies $\phi = 1 + \frac{1}{\phi}$.

We therefore have ϕ as a root of the quadratic $\phi^2 - \phi - 1$. The quadratic formula yields two roots $\frac{1 \pm \sqrt{5}}{2}$. Because ϕ is clearly positive, we must have $\phi = \frac{1+\sqrt{5}}{2}$.

8.10.   We let h_n be the number of moves required to solve the puzzle with n disks. Clearly $h_1 = 1$. For 2 disks, we move the top disk to the right peg, then the bottom disk to the middle peg, then the top disk to the middle peg. Thus $h_2 = 3$. Notice that we just as easily could have used three similar moves to transfer the stack of 2 disks to the right peg instead of to the middle peg. In fact, we require three moves to transfer 2 disks from one peg to any other peg. This idea is important as we consider the general case.

For n disks, we can't move the bottom disk to the middle peg until all the disks on top of it have been moved to the right peg. Because there are $n - 1$ disks on top of the bottom disk, we require h_{n-1} moves to transfer these disks to the right peg. Then we need one move to transfer the largest disk to the middle peg, followed by h_{n-1} moves to transfer the remaining disks to the middle peg. Thus $h_n = 2h_{n-1} + 1$, giving us a recurrence relation for h_n, $n \geq 1$.

There are several ways to solve this recurrence relation. You can use generating functions, but it's easier to iterate for a while and look for a pattern:

$$h_1 = 1$$
$$h_2 = 2 \cdot 1 + 1 = 3$$
$$h_3 = 2 \cdot 3 + 1 = 7$$
$$h_4 = 2 \cdot 7 + 1 = 15$$
$$h_5 = 2 \cdot 15 + 1 = 31$$

We guess that $h_n = 2^n - 1$. This formula can be proven by induction. The base case, $n = 1$, holds. We assume $h_k = 2^k - 1$ and look at h_{k+1}. By the recurrence relation, $h_{k+1} = 2h_k + 1$, which equals $2(2^k - 1) + 1$ by our induction assumption. Thus we have $h_{k+1} = 2(2^k - 1) + 1 = 2^{k+1} - 2 + 1 = 2^{k+1} - 1$, and our result holds.

What about the monks and the end of the world? Each move requires 1 second, so moving 64 disks requires $2^{64} - 1$ seconds. There are about $60 \cdot 60 \cdot 24 \cdot 365 = 31{,}536{,}000$ seconds in a year, so the monks' task will be completed in $(2^{64} - 1)/31{,}536{,}000$ years. This is about 5.85×10^{12} years. So it appears we still have some time to enjoy our planet.

Stepping back

We have a recurrence relation, $s_{n+1} = As_n + Bs_{n-1}$, with initial values s_0 and s_1. From our experience with generating functions, we often found solutions to recurrence relations in which n was an exponent. The simplest such formula for s_n might have the form $s_n = x^n$ for some x. Writing the recurrence relation in the form $s_{n+1} - As_n - Bs_{n-1} = 0$ and substituting this simple formula for s_n, we have $x^{n+1} - Ax^n - Bx^{n-1} = 0$. If we assume x is not 0 (a reasonable assumption if we want a nontrivial sequence), we can divide this equation by x^{n-1} to obtain the quadratic equation $x^2 - Ax - B = 0$. Thus for $s_n = x^n$ to be a solution to our recurrence relation, x must be a root of this quadratic, which is called the *characteristic polynomial* for the recurrence relation.

Now we assume that α_1 and α_2 are distinct roots of the characteristic polynomial. We want to show that there exist constants C_1 and C_2 such that for all $n \geq 0$,

$$s_n = C_1(\alpha_1)^n + C_2(\alpha_2)^n.$$

To find C_1 and C_2, we substitute the values of s_0 and s_1 to obtain a system of two equations in two unknowns:

$$s_0 = C_0 + C_1 \text{ and } s_1 = C_1\alpha_1 + C_2\alpha_2.$$

Because α_1 and α_2 are distinct, the two equations will always have a solution. (The equations are lines in the C_1C_2-plane, and they have different slopes, so they must intersect.) Therefore we can solve for C_1 and C_2 to obtain a formula for s_n. If α_1 and α_2 were not distinct, the lines could be parallel, and this approach would fail. The formula for s_n in the case of a double root α is $s_n = C_1(\alpha)^n + nC_2(\alpha)^n$. We will not explore the derivation of this formula in this text.

In Challenge 8.9, the Fibonacci sequence was given by $F_0 = F_1 = 1$ and $F_n = F_{n-1} + F_{n-2}$ for all $n \geq 2$. The characteristic polynomial of the recurrence relation is $x^2 - x - 1$, which has roots $\frac{1 \pm \sqrt{5}}{2}$. The formula we obtained previously,

$$F_n = \frac{1}{\sqrt{5}}\left(\left(\frac{1+\sqrt{5}}{2}\right)^{n+1} - \left(\frac{1-\sqrt{5}}{2}\right)^{n+1}\right),$$

can be rewritten as

$$F_n = \left(\frac{1+\sqrt{5}}{2\sqrt{5}}\right)\left(\frac{1+\sqrt{5}}{2}\right)^n + \left(\frac{1-\sqrt{5}}{2\sqrt{5}}\right)\left(\frac{1-\sqrt{5}}{2}\right)^n$$

to show the constants $C_1 = \frac{1+\sqrt{5}}{2\sqrt{5}}$ and $C_2 = \frac{1-\sqrt{5}}{2\sqrt{5}}$.

We note that solving for the roots of a characteristic polynomial and then finding constants C_1 and C_2 is often easier than manipulating generating functions. When solving a recurrence relation, if iteration doesn't reveal an obvious pattern, try the characteristic polynomial method next. But don't lose sight of the fact that we know this approach is valid only because of its connection to generating functions.

9

Discrete thoughts of counting
Quantifying complicated quantities

This module presents the fundamental concepts of permutation and combination, along with the Pigeonhole Principle and the Binomial Theorem. These topics are all central in discrete mathematics and are traditionally included in such a course. Except for Challenge 9.3, the material is independent of previous modules.

Challenges 9.3 and 9.4 offer applications of the Pigeonhole Principle that are more subtle than the other Pigeonhole Principle challenges. Note that the proof given to Challenge 9.3 below relies on results from Module 4. If Module 4 is not covered, students could still develop a similar argument using more basic understanding of division and remainders. Alternatively, Challenge 9.3 could be skipped without loss of continuity.

Solutions

9.1. The statement is a theorem. *Extension:* Let N and M be natural numbers such that $N > M > 1$. If N objects are placed in M boxes, then at least one box has $\lceil N/M \rceil$ or more objects in it, where $\lceil N/M \rceil$ denotes the least integer greater than or equal to N/M, also called the *ceiling* of N/M.

Proof: First we note that for any real number a, $a \leq \lceil a \rceil < a + 1$. So we have $\lceil N/M \rceil < N/M + 1$. Now suppose every box has at most $\lceil N/M \rceil - 1$ objects. Then each box contains fewer than N/M objects, giving a total of fewer than $M \times (N/M) = N$ objects, which is a contradiction.

Remark: Ask students if they know why this result is called the Pigeonhole Principle. (Suggest they think of the objects as pigeons and the boxes as pigeonholes.)

We can make a further extension: If an infinite number of objects are placed into a finite number of boxes, then there must be at least one box with an infinite number of objects. The proof follows immediately if we consider how many objects we have if each box contains only a finite number of objects.

9.2.   If we allow for the possibility that someone may have been born on February 29, then there are 366 possible birthdays. Therefore we would need 367 people in a room to guarantee a match. (If we consider only 365 possible birthdays, then the answer is 366 people.)

9.3.   The statement is a theorem. *Extension:* Let $a_0, a_1, a_2, \ldots, a_N$ be integers. There exist distinct indices m and n such that N divides $a_n - a_m$.

Proof: Without the requirement that m and n be distinct, we would just choose $m = n$, and the result holds. Following the hint in Appendix 1, we now look at the remainders obtained when we divide each a_i by N. By Challenge 4.3, there are only N possible remainders: $0, 1, \ldots, N-1$. Thus by the Pigeonhole Principle, there exist distinct m and n such that a_m and a_n have the same remainder when divided by N—that is, they are congruent modulo N. Therefore by Challenge 4.2, N divides $a_n - a_m$.

9.4.   The statement is false. *Counterexample:* Let the four points be the four corners of the square. Then each point is distance at least 2 from the other three points, and the result fails. *Salvage:* The result holds if we require the set S to contain five points.

    Proof: Draw one vertical and one horizontal line through the 2×2 square to divide it into four 1×1 squares. Because there are four squares and five points, by the Pigeonhole Principle, there must be a 1×1 square with at least two points in it. The farthest apart these points could be is $\sqrt{2}$, the length of the diagonal of the small square.

9.5.   The statement is false. There are $(365)^2 = 133{,}225$ possible birthday-deathday pairs. So it is possible in a group of 130,000 that no two people would share both days. *Salvage:* In any group of 133,226 people, there must exist two individuals who share the same birthday (month and day) and will share the same deathday. (If we allow February 29, then the number must be $(366)^2 + 1 = 133{,}957$.)

    Proof: As stated above, there are $(365)^2 = 133{,}225$ possible birthday-deathday pairs. With 133,226 people, at least two must have exactly the same pair of days.

9.6.   There are six permutations of $\{A, E, T\}$: *AET, ATE, EAT, ETA, TAE, TEA*.

9.7.   The statement is a theorem. *Extension:* The natural extension to r-permutations occurs in the next challenge

Proof: From a set of n objects, there are n choices for the first object in the permutation. For each of these choices, there are $n - 1$ choices for the next object, for a total of $n(n - 1)$ so far. For each of these choices, there are $n - 2$ choices for the third object, and so on, until there is only one choice left for the last object. Thus the total number of permutations is $n(n - 1)(n - 2) \cdots (2)(1) = n!$.

Remark: The next challenge introduces the idea of an r-permutation. Given a set S of n objects and a natural number $r \leq n$, an *r-permutation of S* is a permutation of r distinct objects chosen from the set S. The number of r-permutations of a set of n objects is $n(n - 1)(n - 2) \cdots (n - r + 1)$, because we have n choices for the first object, $n - 1$ choices for the second, $n - 2$ for the third, and so on until r

objects have been chosen. The number of r-permutations of a set of n objects is often denoted $P(n, r)$. Note that $P(n, r) = n!/(n - r)!$.

Notice that when $r = n$, the formula $P(n, r) = n!/(n - r)!$ has 0! in the denominator. There is no natural way to interpret 0! as a product, but mathematicians have defined the expression 0! to have value 1. Doing so makes all kinds of formulas work nicely. If we like, we can think of 0! as the number of ways to permute 0 objects. There is only one way to create such a permutation—to do nothing.

9.8.   The answer to the first question is $26 \cdot 25 \cdot 24 \cdot 23$. We have 26 choices for the first letter, 25 for the second, and so on. *ROUNDS* and *CAMPUS* are 6-permutations of the alphabet, because each contains 6 distinct letters. *BURGER* is not a 6-permutation, because it contains a repeated letter. The number of r-permutations of a set of n objects is $n(n - 1)(n - 2) \cdots (n - r + 1)$, because we have n choices for the first object, $n - 1$ choices for the second, $n - 2$ for the third, and so on until r objects have been chosen. (See *Remark* after Challenge 9.7.)

9.9.   The answer is $n(n - 1)!(n - 1)!$. First note that because no sophomores may sit together and because there are an equal number of seniors and sophomores, the seniors and sophomores must alternate around the table. Because we are only considering relative position, let's count the seating arrangements from the point of view of one senior, who we will call Julia. To her left must be a sophomore, for whom there are n choices. In the next seat to the left, there must be a senior, for whom there are $n - 1$ choices. Moving left again, there is another sophomore, for whom there are $n - 1$ choices, and so on. The total is $n(n - 1)(n - 1)(n - 2)(n - 2) \cdots (2)(2)(1)(1) = n(n - 1)!(n - 1)!$.

9.10.   *Proof:* We will use the notation and terminology following the challenge. We know from Challenge 9.8 that the number of m-permutations of a set of n objects is $n(n - 1)(n - 2) \cdots (n - m + 1) = n!/(n - m)!$. One way to think of counting all these m-permutations is to first count the number of m-combinations taken from a set of n objects and then note that each such combination can be permuted in $m!$ ways. Therefore we have

$$\binom{n}{m} \cdot m! = \frac{n!}{(n - m)!}.$$

Dividing both sides by $m!$ gives the result.

Remark: It's helpful to point out how this formula makes sense for $m = n$ and $m = 0$. When $m = n$, we are counting the number of ways to choose all n objects, so there is only one way to do it. The formula becomes $\binom{n}{n} = n!/n!0!$, which equals 1 given our definition of $0! = 1$. (See *Remark* after Challenge 9.7.) When $m = 0$, we are counting the number of ways to choose no objects from a set of n objects. There is only one way to do this—to do nothing—and the formula again yields $\binom{n}{0} = n!/0!n! = 1$.

We note finally that all the integers in these formulas for permutations and combinations are nonnegative. There are ways to generalize the binomial coefficients to include negative integers and even all real numbers, but we will not go there in this text.

9.11.   The number of 5-card hands from a 52-card deck is $\binom{52}{5} = \frac{52!}{5!47!}$. The number of 5-card hands that are all hearts is $\binom{13}{5} = \frac{13!}{5!8!}$. Because there are four possible suits from which to choose a flush, the number of flushes is $4 \cdot \frac{13!}{5!8!}$.

Remark: Encourage your students to be wise if they actually want to expand $\binom{n}{m}$ for large values of n or m. If they are using a system like Mathematica, then there should be no cause for concern. But using an ordinary calculator is another matter. For example, if they begin their expansion of $\frac{52!}{5!47!}$ by expecting their calculator to compute 52!, they will get, at best, an approximation; at worst, an error message. Instead, they can cancel the obvious common factors, reducing the expression to $\frac{52(51)(50)(49)(48)}{5!}$ before calculating the integer value.

9.12.   *Proof:* These identities are easy to prove using formulas and algebra. The theorem in Challenge 9.10 gives us

$$\binom{n}{r} = \frac{n!}{r!(n-r)!} \quad \text{and} \quad \binom{n}{n-r} = \frac{n!}{(n-r)!(n-(n-r))!},$$

which are clearly equal. For the second identity, we have

$$\binom{n}{r-1} + \binom{n}{r} = \frac{n!}{(r-1)!(n-r+1)!} + \frac{n!}{r!(n-r)!} = \frac{(r)n!}{r!(n-r+1)!} + \frac{(n-r+1)n!}{r!(n-r+1)!}$$

$$= \frac{(r+n-r+1)n!}{r!(n-r+1)!} = \frac{(n+1)!}{r!(n+1-r)!} = \binom{n+1}{r}.$$

Remark: We note that the first identity holds for any r, $0 \le r \le n$, whereas the second identity requires that $1 \le r \le n$. In the second identity, therefore, we also have $n \ge 1$.

The two identities can also be proved using what is called a *combinatorial argument* or *counting argument*. For such an argument, you specify something you will count in two ways: a set of objects or the number of ways to do something. One way of counting gives the expression on one side of the identity, the other way of counting gives the expression on the other side of the identity. Because you are counting the same thing, the two expressions must be equal.

For the first identity, we count the number of ways to choose r objects from a set of n objects. The straightforward way yields "n choose r" $= \binom{n}{r}$. But notice that for each choice of r objects, you are also, in effect, "choosing" the $n - r$ objects left behind. The number of ways to make this "complementary" choice is $\binom{n}{n-r}$; so the identity holds.

For the second identity, we will count the number of ways to choose r objects from a set of $n + 1$ objects. The straightforward way yields $\binom{n+1}{r}$. To count these choices in another way, we pick one object from the $n + 1$ objects and call it x. (Note that $n \geq 1$, so we know there is such an element x.) Our selection of r objects will either include x or it won't. If our selection includes x, then we have n objects left from which to choose the remaining $r - 1$ objects, for a total of $\binom{n}{r-1}$ ways. If our selection does not include x, then we have to choose all our r objects from the remaining n in our set, for a total of $\binom{n}{r}$. Because each possible selection is counted once in one or the other of these expressions, we have the total number of choices equal to the sum $\binom{n}{r-1} + \binom{n}{r}$. Therefore the identity holds.

9.13. Below are the first seven rows of Pascal's triangle, left-justified for convenience in this challenge. We see that the first entry in each row is $\binom{n}{0}$, which equals 1. In particular, this completes the first row. We can use the identity in Challenge 9.12 to generate each remaining entry as follows: Each entry will be the sum of the entry directly above it and the entry above and to the left. Notice that the rightmost entry in each row has no entry above it. In that case, we just add 0 to the entry above and to the left.

1						
1	1					
1	2	1				
1	3	3	1			
1	4	6	4	1		
1	5	10	10	5	1	
1	6	15	20	15	6	1

Finally, look at the diagonals that start with 1 along the left side and move up at a 45° angle ("northeast"). For each such diagonal, sum the entries: 1, 1, $1 + 1 = 2$, $1 + 2 = 3$, $1 + 3 + 1 = 5$, $1 + 4 + 3 = 8$, and so on, to obtain the Fibonacci sequence!

9.14. *Proof:* Following the hint in Appendix 1, we will use induction. Our base case $N = 0$ holds as follows:

$$(a+b)^0 = 1 \text{ and } \sum_{n=0}^{0} \binom{0}{n} a^n b^{0-n} = \binom{0}{0} a^0 b^0 = 1.$$

Now we suppose the theorem holds for N and show it must therefore hold for $N + 1$.

$$(a+b)^{N+1} = (a+b)(a+b)^N = (a+b)\sum_{n=0}^{N}\binom{N}{n}a^n b^{N-n} \qquad \text{(by induction hypothesis)}$$

$$= a\sum_{i=0}^{N}\binom{N}{i}a^i b^{N-i} + b\sum_{j=0}^{N}\binom{N}{j}a^j b^{N-j} \qquad \text{(distribute }(a+b)\text{)}$$

$$= \sum_{i=0}^{N}\binom{N}{i}a^{i+1} b^{N-i} + \sum_{j=0}^{N}\binom{N}{j}a^j b^{N-j+1} \qquad \text{(multiply through)}$$

$$= a^{N+1} + \sum_{i=0}^{N-1}\binom{N}{i}a^{i+1} b^{N-i} + \sum_{j=1}^{N}\binom{N}{j}a^j b^{N-j+1} + b^{N+1} \qquad \text{(pull out two terms)}$$

$$= a^{N+1} + \sum_{j=1}^{N}\binom{N}{j-1}a^j b^{N-j+1} + \sum_{j=1}^{N}\binom{N}{j}a^j b^{N-j+1} + b^{N+1} \qquad \text{(let } j = i+1 \text{ in first sum)}$$

$$= a^{N+1} + b^{N+1} + \sum_{j=1}^{N}\left(\binom{N}{j-1}+\binom{N}{j}\right)a^j b^{N-j+1} \qquad \text{(combine summations)}$$

$$= a^{N+1} + b^{N+1} + \sum_{j=1}^{N}\binom{N+1}{j}a^j b^{N+1-j} \qquad \text{(by the recurrence formula)}$$

$$= \sum_{j=0}^{N+1}\binom{N+1}{j}a^j b^{N+1-j}. \qquad \text{(bring all terms into sum)}$$

Thus the theorem holds for $N + 1$, and therefore, by mathematical induction, the Binomial Theorem holds for all $N \geq 0$.

Alternate proof: Here is a combinatorial proof: We need to show that for each $n = 0, 1, \ldots, N$, when $(a + b)^N$ is expanded, the term $a^n b^{N-n}$ occurs $\binom{N}{n}$ times. First we observe that when $(a + b)^N$ is expanded, each term is generated by taking either a or b from each of the N factors:

$$(a+b)^N = \overbrace{(a+b)(a+b)(a+b)\cdots(a+b)}^{N}.$$

The term $a^n b^{N-n}$ will occur each time we take a from n of the factors and b from the remaining $N - n$ factors. There are exactly $\binom{N}{n}$ ways to choose the n factors from which to take a. Therefore the term $a^n b^{N-n}$ will occur exactly $\binom{N}{n}$ times, proving the theorem.

Remark: The second argument is a great example of the power of a combinatorial proof. It's much shorter and more intuitive than induction. It also reinforces the fact that the binomial coefficient $\binom{N}{n}$ counts the number of ways to choose n objects from a set of N objects. On the other hand, the induction argument uses the elegant recurrence formula and requires students to manipulate summations, a potentially valuable skill.

9.15.   If we let $a = b = 1$ in the Binomial Theorem, we obtain the identity $\sum_{n=0}^{N} \binom{N}{n} = 2^N$. Now let S be a set with N elements and recall that the power set of S has 2^N elements (see Challenge 6.13). We can now give a combinatorial proof of the identity as follows: Because the power set of S is the set of all subsets of S, S has 2^N subsets. Now count those subsets in a different way. Each subset has either 0 elements, or 1 element, or 2, ... , or N elements. For each $n = 0, 1, 2, \ldots, N$, there are exactly $\binom{N}{n}$ subsets of size n. Therefore the total number of subsets of S is $\sum_{n=0}^{N} \binom{N}{n}$. The identity follows.

Stepping back

For this challenge, we center the rows of the triangle. Notice that the recurrence formula for computing each entry from the row above now applies as follows: After the first row, each entry is the sum of the two entries above it—one above and a little to the left, the other above and a little to the right, where addition is computed mod 2 and an "empty summand" is taken to be 0. The resulting figure resembles the structure of a famous fractal, the Sierpinski Triangle. Smaller triangles have been outlined in the figure to highlight this structure.

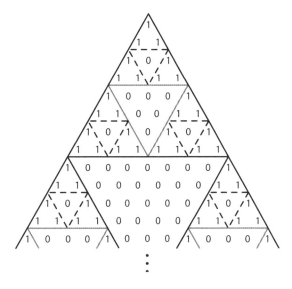

How does the pattern arise? Start with the small triangle of three 1's at the top and think about how the recurrence formula will generate Row 3. The first entry is a 1 (as always), the next entry is a 0, and the last entry is a 1. Now in Rows 3 and 4, the 1 at each end of Row 3 will give rise to a copy of the original small triangle. In particular, Row 4 will consist of all 1's. But now for Row 5, we are in exactly the analogous situation as we were for Row 3. Following a row of all 1's, the recurrence formula will always yield a row that begins and ends with 1 with all 0's in between. Notice the larger triangle pattern that has been created in the first four rows. Because Row 5 has only 0's in the middle entries, the 1's at each end will generate a copy of the larger, 4-row triangle, ending in a row of all 1's at Row 8. Row 9 will then arise with a 1 at each end and 0's in between, and the pattern continues. The self-replicating quality of this pattern reflects one of the most fascinating characteristics of fractals.

10

Quantifying uncertainty with probability

A likely story?

Module 10 introduces basic probability on finite sets in which every outcome is equally likely. Early challenges develop the basic properties of a finite, equiprobable space; later challenges offer classic probability scenarios involving coin tosses, playing cards, and the ever popular matching-birthday question. This module is a natural follow-up to Chapter 9. In particular, Challenges 10.5 and 10.9 make use of binomial coefficients. The other challenges require some very basic counting that could be developed independently of Chapter 9. Note that throughout this module we assume our sample space S is nonempty.

Solutions

10.1.   The statement is false. *Counterexample:* Let $A = B$. Then $P(A \cup B) = P(A)$, but $P(A) + P(B) = 2P(A)$. *Salvage:* If A and B are events in $\mathcal{P}(\{s\})$, then $P(A \cup B) = P(A) + P(B) - P(A \cap B)$.

Proof: By definition, $P(A \cup B) = \frac{|A \cup B|}{|S|}$, so we need to show that $|A \cup B| = |A| + |B| - |A \cap B|$. Notice that $|A| + |B|$ counts all the elements of $A \cap B$ twice. Thus if we subtract the number of elements in $A \cap B$, each element in the union will be counted exactly once. Therefore $|A \cup B| = |A| + |B| - |A \cap B|$, and the result holds.

10.2.   We apply the definition of the probability function to verify each property.

- $\mathcal{P}(\{s\}) = |S|/|S| = 1$
- $\sum_{s \in S} \mathcal{P}(\{s\}) = \sum_{s \in S} 1/|S| = |S|/|S| = 1$
- $P(\bar{A}) = |\bar{A}|/|S| = (|S| - |A|)/|S| = 1 - |A|/|S| = 1 - P(A)$
- $P(\emptyset) = |\emptyset|/|S| = 0$

10.3.   Applying the definition yields the results.

- By definition, $P(A) = |A|/|S|$. Because A is a set, $|A| \geq 0$; because S is nonempty, $|S| > 0$; therefore $P(A) \geq 0$.

- Because all sets under consideration are finite, we know that $A \subseteq B$ implies $|A| \leq |B|$. Therefore $|A|/|S| \leq |B|/|S|$, and so $P(A) \leq P(B)$.

- Because all sets are finite, for any $A \subseteq S$, we have $0 \leq |A| \leq |S|$. Therefore $0 \leq |A|/|S| \leq 1$, so $0 \leq P(A) \leq 1$.

10.4.   The answer is "No." When the two numbers are first selected, the resulting fraction falls into one of four categories: odd/odd, odd/even, even/even, and even/odd. Because half the numbers from 1 to 1,000,000 are even and half are odd, each of these four categories is equally likely. Therefore, before being reduced to lowest terms, half of the possible fractions have the form odd/even or even/odd, which will never be reduced to a fraction of the form odd/odd. In addition, there are fractions of the form even/even that do not reduce to odd/odd, such as 4/6 = 2/3. Therefore more than half of all possible fractions will not reduce to the form odd/odd.

10.5.   Let S be the sample space for the experiment of flipping a fair coin 10 times. Then $|S| = 2^{10} = 1024$. Let A be the event "exactly five heads." Then $|A|$ will equal the number of ways in which exactly 5 of the 10 flips come up heads, which is $\binom{10}{5} = 252$. Therefore $P(A) = 252/1024 \approx 0.246$.

To have exactly 5 heads in a row, we observe that the first such head must occur on flip 1, 2, 3, 4, 5, or 6. The remaining flips in such a sequence would be completely determined. Therefore the probability of exactly 5 heads in a row is $6/1024 \approx 0.00586$.

The probability of seeing at least 1 head is easy to determine if we consider the complementary event. We note that $P(\text{no heads}) = P(\text{all tails}) = 1/1024$. Therefore by the third property in Challenge 10.2, we have $P(\text{at least 1 head}) = 1 - 1/1024 \approx 0.999$.

10.6.   The probability that all three coins land with the same face up is 0.25.

Proof: Let S be the sample space for the experiment of flipping three different fair coins. Then $|S| = 2^3 = 8$. Because there is only one way for the coins to land with all heads up and only one way with all tails up, we have $P(\text{all heads or all tails}) = 2/8 = 0.25$.

The reasoning that leads to an answer of .5 as suggested in the text is faulty. There are six ways for two coins to show the same face with the third coin showing the opposite face: HHT, HTH, THH, TTH, THT, HTT. Therefore six out of eight tosses will *not* yield three matching coins. The faulty argument does not take into account the order in which the heads and tails occur.

10.7.   Let S be the sample space for the experiment of choosing 10 cards in succession from a deck of 52, replacing each card after it is chosen. Then $|S| = 52^{10}$. Let A be the event in which all 10 cards are different. So each element of A is a permutation of 10 cards from the deck of 52, giving us $|A| = 52 \cdot 51 \cdot \cdots \cdot 43$. Therefore the probability of 10 cards with no match is

$$P(A) = \frac{52 \cdot 51 \cdot \cdots \cdot 43}{52^{10}} = \frac{52}{52} \cdot \frac{51}{52} \cdots \frac{43}{52} \approx 0.397.$$

Thus we have that the probability of getting at least one match in 10 cards is approximately $1 - 0.397 = 0.603$.

10.8.   The probability that the other side of the card is blue is 1/3. Here's one way to arrive at this answer: Let the sample space S be the six sides of the cards. Thus S has three red elements and three blue elements. If we are shown a red side, it could be any one of the three. But two of these three sides are on a card with red on the other side. Therefore the probability that we are looking at the card with blue on the other side is 1/3.

10.9.   Let S be the sample space for the experiment of dealing 5 cards from a standard 52-card deck. Then $|S| = \binom{52}{5} = 2{,}598{,}960$. The number of hands with 4 aces equals the number of ways to choose the 5th card from those remaining, which is 48. So $P(4 \text{ aces}) = 48/\binom{52}{5} \approx 0.00001847$.

A flush is a hand in which all 5 cards are from the same suit. There are four ways to choose the suit and then $\binom{13}{5}$ ways to choose the 5 cards, for a total of $4\binom{13}{5}$ possible flushes. So $P(\text{flush}) = 4\binom{13}{5}/\binom{52}{5} \approx 0.001981$.

A full house is a hand with three of one kind and two of another. There are 13 choices for the first kind, with $\binom{4}{3}$ ways to choose 3 cards of that kind. Then there are 12 choices for the second kind, with $\binom{4}{2}$ ways to choose two cards of that kind. So the number of full houses is $13\binom{4}{3} \cdot 12\binom{4}{2}$. So $P(\text{full house}) = 13\binom{4}{3} \cdot 12\binom{4}{2}/\binom{52}{5} \approx 0.00144$.

10.10.   The probability that 2 or more of the 40 people share a birthday is 1 minus the probability that no two share a birthday. Following the argument in Challenge 10.7, we find that the probability of no shared birthdays is

$$\frac{365}{365} \cdot \frac{364}{365} \cdot \frac{363}{365} \cdots \frac{326}{365} \approx 0.108768.$$

Therefore the probability that 2 of your 40 people share a birthday is approximately $1 - 0.108768 = 0.891232$. So more than 89% of the time, in a group of 40 people, there will be 2 who share a birthday. Pretty amazing!

To answer the second question, we use the same technique to compute the probability of a shared birthday among groups of size n for various sizes of n. We find smallest n for which the probability is greater than 1/2 is $n = 23$.

Stepping back

To analyze this game carefully, we need to explore the idea of *expected payoff*—that is, the average payoff received if the game were to be played many times. Suppose a very simple game had two possible payoffs, $2 and $4, each with probability of 1/2. If the game were played many times, we would expect a payoff of $2 half the time and $4 the other half, for an average payoff of $\frac{\$2 + \$4}{2} = \$3$. Notice that this average can also be obtained as follows:

$$Expected\ payoff = (1/2)\$2 + (1/2)\$4 = P(payoff = \$2) \cdot \$2 + P(payoff = \$4) \cdot \$4 \,.$$

This leads to the definition of *expected payoff* (or *expected value*) $EV(G)$ of a game G with possible payoffs $p_1, p_2, \ldots, p_n$:

$$EV(G) = \sum_{k=1}^{\infty} P(payoff = p_k) \cdot p_k \,.$$

A game is called *fair* if the price to play the game equals the expected payoff. Thus for the simple game above with expected payoff $3 to be fair, we should pay $3 each time we play the game.

For the St. Petersburg Paradox, we need to determine the probabilities for each possible payoff. The $2 payoff occurs when the first toss is heads, which occurs with probability 1/2. The payoff is $4 when the first two tosses come up TH. This outcome occurs with probability 1/4 because it is one of four equally likely outcomes. The payoff is $8 when the first three tosses come up TTH. This outcome occurs with probability 1/8 because it is one of eight equally likely outcomes. In general, the payoff is $\$2^n$ when the first n tosses come up TT$\cdots$ TH, with $n - 1$ T's, which occurs with probability $1/2^n$. Extending our definition of expected payoff to an infinite sum, we find the expected payoff of the St. Petersburg Paradox G to be

$$EV(G) = \sum_{n=1}^{\infty} P(payoff = \$2^n) \cdot \$2^n = \sum_{n=1}^{\infty} (1/2^n) \cdot \$2^n = \sum_{n=1}^{\infty} \$1 \,.$$

Yikes! The average payoff is infinity dollars. Technically, to be fair, you should be willing to pay infinity dollars to play this game!

Let's see what happens if you pay $1000 to play this game. To make money, you need a payoff greater than $1000, or at least $2^{10} = \$1024$. So you need to toss at least nine tails before you get your first head. The probability of tossing tails nine times in a row is $1/2^9 = 1/512 \approx 0.00195$. Thus the chances that you'll make money are very small. The paradox is that even though the average payoff for this game is infinity dollars, the odds that you'll make money by playing the game are extremely small.

11

The subtle art of connecting the dots
Edging up to graphs

Module 11 introduces notions from elementary graph theory, including Euler circuits, Euler paths, complete graphs, and a brief introduction to trees. This material is fundamental and standard fare in discrete mathematics. It also opens the door for further study in graph theory, a field in which there are many interesting questions accessible to undergraduates.

Solutions

11.1. The statement is a theorem. *Extension:* The total degree of a graph equals twice the number of edges.

Proof: When we sum the degrees of the vertices, an edge joining distinct vertices is counted twice. Loops are also counted twice by our convention. Thus each edge is counted exactly twice, and the result follows.

11.2. The statement is a theorem. There is no natural extension.

Proof: If a graph had an odd number of vertices of odd degree, then the total degree of the graph would be odd as well, contradicting Challenge 11.1.

11.3. The first graph exists as shown in the figure. The remaining two graphs do not exist because each has an odd number of vertices of odd degree.

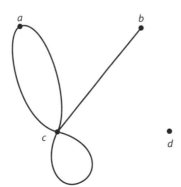

11.4.   The statement is a theorem. There is no natural extension.

Proof: By the definition of connected, for any two vertices a and b, there is a path from a to b. Let P be a path from a to b with the least number of edges. Suppose P has a repeated vertex, v so that P has the form: $a, e_1, v_1, e_2, \ldots, e_i, v, e_{i+1}, \ldots, v, e_j, \ldots, b$. Then the path $a, e_1, v_1, e_2, \ldots, e_i, v, e_j, \ldots, b$ has fewer edges. This contradiction implies that P has no repeated vertices and, thus, is a simple path.

11.5.   The statement is a theorem. *Extension:* Let a and b be vertices on a circuit C of G. If an edge or a vertex not equal to a or b is removed from C, then the resulting graph still contains a path from a to b.

Proof: There are clearly at least two paths from a to b in G: Start at a and traverse the circuit in one direction to reach b or traverse the circuit in the other direction to reach b. Because a circuit has no repeated edges, these two paths are edge-disjoint, that is, they have no edges in common. Thus removing a single edge of C will leave at least one path from a to b intact. Similarly a circuit has no repeated vertices, so the two paths above have no vertices in common except for a and b. Thus removing a vertex not equal to a or b will leave one of the two paths intact.

Remark: Note that the argument above is rigorous without introducing notation for individual vertices and edges in the circuit. Such notation is particularly awkward in this proof because we don't know where a and b lie in the circuit. Encourage students to avoid excessive notation if they can.

11.6.   The statement is false. Removing edge e from the graph below will disconnect the graph. *Salvage:* If G is connected and contains a circuit C, then removing an edge of C yields a connected graph.

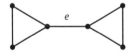

Proof: Let the edge e removed from C have endpoints a and b. Let u and v be arbitrary vertices of G. If a path from u to v in G does not include e, then that path remains in $G \setminus \{e\}$. If a path from u to v in G includes e, then, by Challenge 11.5, removing e leaves an alternative path between a and b. This alternative path can be used as a detour to successfully navigate from u to v.

11.7.   The statement is a theorem. *Extension:* See Challenge 11.9.

Proof: If G has an Euler circuit, then the result follows easily. The Euler circuit contains every vertex, so a walk from any vertex to any other vertex can be found merely by following the circuit. In addition, to traverse the Euler circuit, every edge used to "enter" a vertex must be paired with a second edge used to "exit" the vertex. Because no edge may be used more than once, the edges at any particular vertex must pair up, giving an even degree.

Now suppose *G* is connected and every vertex has even degree. If *G* has only one vertex, the result holds trivially, so suppose *G* has at least two vertices. Consider a path *P* with the maximum possible number of edges starting at some vertex *v*. While traversing *P*, every edge used to "enter" a vertex can be paired with a unique second edge used to "exit" that vertex, because each vertex has even degree and no edge can be repeated. Thus edges occur in entrance-exit pairs. Because *G* is finite, *P* must have a final edge, which must be paired with the initial edge of *P* used to exit the starting vertex *v*. Thus *P* is, in fact, a circuit. Now suppose there is some edge *e* not contained in *P*. Then starting with *e*, we can construct another circuit following the same method as that for *P*. But then this new circuit can be appended to *P* to create a longer path, which contradicts the choice of *P*. Thus *P* contains all the edges of *G*. Having all edges in *P* implies all vertices are in *G* because *G* is connected. Therefore *P* is an Euler circuit for *G*.

11.8.   The statement is true. *Extension:* If a vertex of *G* has odd degree or if *G* is not connected, then *G* does not contain an Euler circuit.

Proof: This result follows immediately from Challenge 11.7.

11.9.   The statement is false. Here's a graph with two vertices of odd degree and no Euler path:

Salvage: The graph *G* has an Euler path if and only if it is connected and has zero or two vertices of odd degree.

Proof: If *G* has an Euler path, then *G* is clearly connected. If the path is a circuit, then applying the entrance-exit pairing of edges used in Challenge 11.9 yields all even degree vertices; thus *G* has no vertices of odd degree. If the path is not a circuit, then suppose it starts at vertex *a* and ends at *b*. Add an edge between *a* and *b*, which now allows the creation of an Euler circuit in the larger graph. By Challenge 11.8, this graph has all vertices of even degree. Removing the added edge leaves exactly two vertices of odd degree: *a* and *b*.

To prove the converse in the zero odd degree case, apply Challenge 11.9. In the case where two vertices have odd degree, just add an edge between them, apply Challenge 11.9, and then delete the extra edge to get an Euler path.

11.10.   It is not possible to stroll across each bridge in Königsberg exactly once and return to the starting point. Modeling the town with a graph where each landmass is a vertex and each bridge is an edge yields a graph with four vertices of odd degree.

11.11.   The statement is a theorem. *Extension:* Any tree with more than one vertex has at least two vertices of degree 1. (Such vertices are called *leaves*.)

Proof: Let T be a tree with at least two vertices. Let P be a path with the maximum number of edges. Because T has no circuits, P will have no repeated vertices. Thus the first vertex and the last vertex of P will have degree 1. Otherwise P could be made longer.

11.12.   The statement is false. The first graph in the figure in the text is a tree with 25 vertices and 24 edges. *Salvage:* A graph G with n vertices is a tree if and only if G is connected and has $n-1$ edges.

Proof: Suppose G is a tree. Then G is connected by definition. We use induction on n to prove the result. A tree with one vertex has no edges. Suppose all trees with $k \geq 1$ vertices have $k-1$ edges. Consider a tree T with $k+1$ vertices. By Challenge 11.11, T has at least one leaf. Delete one leaf. The result is still connected, has no circuits, and thus is a tree with k vertices. By our induction hypothesis, this tree has $k-1$ edges. Reinserting the leaf and its edge yields the result.

Now suppose G is connected and has $n-1$ edges. We need to show G is circuit-free. Suppose G has a circuit C. By Challenge 11.6, we can delete an edge of C without disconnecting G. If the resulting graph still has a circuit, we repeat the process. We continue until the final graph has no circuits. It will still be connected and will have fewer than $n-1$ edges. But a connected graph with no circuits is a tree, and we showed above that a tree must have $n-1$ edges. This contradiction shows that our original graph must have been circuit-free and therefore is a tree.

11.13.   The statement is a theorem. *Extension:* If a and b are distinct vertices of a tree G, then there is a unique simple path between a and b.

Proof: G is connected, so by Challenge 11.4, there is at least one simple path from a to b. If there are two simple paths, then as we leave a on our way to b, at some vertex the two paths must diverge. When they come back together again at or before b, we will have a circuit, which cannot exist in G. Therefore the result holds.

11.14.

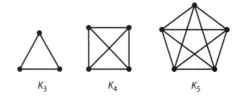

K_3 K_4 K_5

11.15.   The statement is false. K_3 has three vertices, but $[n^2/2] = 4$. *Salvage:* The graph K_n contains $n(n-1)/2$ edges.

Proof: Each vertex in K_n has degree $n-1$, giving a total degree of $n(n-1)$. By the extension of Challenge 11.1, we know $n(n-1)$ equals twice the number of edges in the graph. Therefore the result holds.

Stepping back

Proof: Because G has no loops and at most one edge between each pair of vertices, G will have at most as many edges as K_n. The result follows from the salvage of Challenge 11.15.

12

Just plane graphs
Drawing without being cross

$\mathbf{M}$odule 12 introduces planar graphs and the idea of isomorphism. Early challenges develop the elegant Euler characteristic for connected planar graphs and then use it to prove that $K_{3,3}$ and K_5 are not planar. The ideas of isomorphic and homeomorphic graphs follow, leading to a statement (without proof) of Kuratowski's powerful theorem characterizing all planar graphs. The module ends with the special case of regular planar graphs, laying the groundwork for the proof in Module 13 that there are only five regular polyhedra (the Platonic solids). This is excellent material for a course on discrete math, as well as for courses on proof or higher mathematics.

Solutions

12.1. The statement is a theorem. *Extension:* Let G be a circuit-free graph. Then G is planar.

Proof: A circuit-free graph is just a collection of disjoint trees (called a *forest*). It suffices to prove that a tree is planar. Start with one vertex and draw all incident edges outward. Continue drawing edges in larger, concentric layers. Because a tree has no circuits, no edges need cross.

12.2. The statement is a theorem. *Extension:* If G is a planar graph with m "pieces" (called *connected components*; see the *Remark* below), then the Euler characteristic of G is $m + 1$.

Proof of original statement: Let G be a connected planar graph. We proceed by induction on the number of circuits in G. If G has no circuits, then G is a tree. By Challenge 11.12, G has $V - 1$ edges, and any planar drawing of G divides the plane trivially into one face. Thus $V - E + F = 2$. Now consider a connected planar graph with k circuits, $k \geq 1$. In a planar drawing of G, look at one circuit, call it C. Independently of all other circuits in the drawing, C divides the plane into two regions, an inside and an outside. (This follows from the very deep Jordan Curve Theorem.) Now remove an edge from C. This will collapse two regions into one. By Challenge 11.6, the resulting graph is still connected and is obviously still planar, but it has one fewer edge than G, one fewer region, and one fewer circuit. By induction, this graph has an Euler characteristic of 2. Adding back the edge and thereby increasing the number of regions by 1 gives an Euler characteristic of 2 for G as well.

Proof of extension: Let G be a planar graph with m connected components. Then for the ith component of G, we have the equation $V_i - E_i + F_i = 2$. Adding all these equations gives $2m$ on the right side. But on the left side we have counted the unbounded face m times, so $V - E + F$ must equal $2m - (m - 1) = m + 1$, as desired.

Remark: To define a connected component, we first recall that a *subgraph* of a graph G is a graph with vertices and edges contained in G. A subgraph H of G is a *connected component* if it is connected and is not properly contained in any other connected subgraph of G—that is, H is a maximal connected subgraph of G. In the example below, G has many subgraphs but only two components.

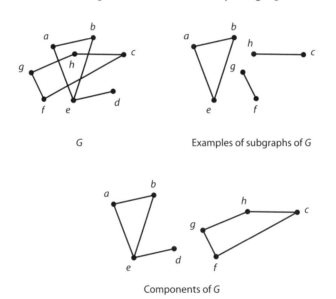

G   Examples of subgraphs of G

Components of G

12.3.   All appear to be planar except K_5 and $K_{3,3}$.

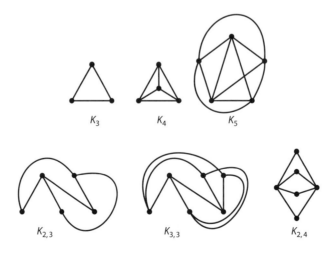

12.4.   The statement is false. *Counterexample:* Consider K_3 with a loop added at each vertex. *Salvage:* Let G be a connected, planar graph having V vertices and E edges, with no loops or multiple edges. If $V \geq 3$, then $E \leq 3V - 6$.

Proof: Because G has no loops or multiple edges, every region in a planar drawing of G will be bounded by at least three edges. Now imagine walking around the boundary of each region, counting edges we walk along as we go. If an edge is part of a circuit, it will be counted exactly twice—once as we walk along the region inside the circuit, and once as we walk along the region outside the circuit. If an edge is not part of a circuit, it must somehow stick out into the interior of a region. Thus we will count such an edge twice as well, walking up one side, around the vertex at the tip, and down the other side. Summing over all the regions, we thus get $2E$. But because each region is bounded by at least three edges, we deduce that $2E \geq 3F$.

Now use the fact that G has Euler characteristic 2. Thus we have $3V - 3E + 3F = 6$. Into this substitute the inequality $2E \geq 3F$ to get $3V - 3E + 2E \geq 6$. The result follows.

Remark: That $2E \geq 3F$ for graphs satisfying the hypothesis of the salvaged statement could be stated as a lemma for your students to prove separately.

12.5.   To show K_5 is not planar, observe that K_5 has 5 vertices and 10 edges and thus does not satisfy the inequality in Challenge 12.4.

12.6.   The statement is a theorem. *Extension:* This result holds for more general bipartite graphs, as long as they are planar, connected, and have no multiple edges. (See *Remark* below.)

Proof of original statement: First we show that if G is a complete bipartite graph, then any circuit in G contains at least four edges. Because the edges of G must join vertices in two disjoint subsets, any circuit in G must alternate vertices from these two sets. In addition, because G has no multiple edges, a circuit must contain at least four vertices and therefore at least four edges.

Now we use the same approach used for Challenge 12.4, noting that now every region in a planar drawing of G will have at least four edges on its boundary. Thus we obtain the inequality $2E \geq 4F$. Substituting this inequality into the Euler characteristic formula $2V - 2E + 2F = 4$ gives $2V - 2E + E \geq 4$, and the result follows.

Remark: A graph G is bipartite if its vertex set can be partitioned into two, nonempty, disjoint sets of vertices V_1 and V_2, so that every edge in G joins a vertex in V_1 to a vertex in V_2. Sometimes it is helpful to think of the vertices in the two sets as being of two different colors. Then all edges in the graph must join vertices of different colors, as in the examples below.

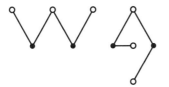   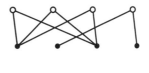

If you give your students this definition (or they discover it for themselves), you can have them use the approach above to show that any circuit in a bipartite graph must have an even number of edges. If the graph has no multiple edges, then all circuits have even length at least 4.

12.7.   To show $K_{3,3}$ is not planar, observe that $K_{3,3}$ has six vertices and nine edges and, thus, does not satisfy the inequality in Challenge 12.6. For the *Gas, Water, Electric Puzzle*, observe that the scenario can be modeled with a graph isomorphic to $K_{3,3}$: Each of three houses is joined by an "edge" to each of three utilities. Because $K_{3,3}$ is not planar, it is impossible to draw the utility lines in the plane without at least two lines crossing each other.

12.8.   The statement is a theorem. *Extension:* W_n is isomorphic to W_m if and only if $n = m$.

Proof: The extension here is minimal. If $n = m$, of course W_n is isomorphic to W_m. For the other direction, we prove the contrapositive. If $n \neq m$, then W_n and W_m have different numbers of vertices (and edges), so they can't be isomorphic.

12.9.   The statement is false. *Salvage:* The two graphs in the figure are not isomorphic.

Proof: Often, there are many ways to show two graphs are not isomorphic. One way, in this case, is to note that the graph on the left has a circuit with three edges whereas the graph on the right does not.

Remark: You might ask your students to what familiar graph the graph on the right is isomorphic. The answer is $K_{3,3}$.

12.10.   The statement is a theorem. *Extension:* If G has a subgraph isomorphic to $K_{3,3}$ or to K_5, then G is not planar.

Proof: If G were planar, then a planar drawing would contain a planar drawing of $K_{3,3}$ or K_5.

12.11.   The statement is false. *Counterexample:* Consider the graph H obtained from $K_{3,3}$ by adding a single vertex in the middle of one edge. H itself is clearly not isomorphic to $K_{3,3}$, nor does it contain a subgraph isomorphic to $K_{3,3}$ or to K_5. (Make sure the students see why.) Yet H is not planar. If it were, a planar drawing would yield a planar drawing of $K_{3,3}$ through removal of the additional vertex.

Remark: Attempts to salvage this statement will lead to the idea of homeomorphism. Maybe your students will develop this idea entirely on their own, or they might discover it by reading ahead in the text, which is why we don't state a formal extension here.

12.12.   The statement is a theorem. *Extension:* The graph contains two subgraphs homeomorphic to K_4.

Proof: See the figure.

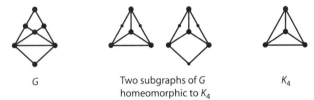

G   Two subgraphs of G K_4
                     homeomorphic to K_4

Remark: The figure illustrates a way to draw homeomorphic copies of a graph. Reducing the size of the vertices from G that were "added" to edges of K_4 helps clarify the homeomorphic relation.

12.13.   Here is a subgraph of the Peterson graph that is homeomorphic to $K_{3,3}$.

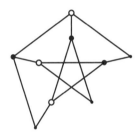

Remark: Note here that the "added'" vertices are again small. In addition, the two sets of vertices in $K_{3,3}$ are colored black and white to make the graph structure easier to see.

12.14.   The first graph has a symmetric planar drawing, as the figure illustrates: Each vertex has degree 2 and each of the two regions is bounded by two edges. The third graph is also shown with a symmetric planar drawing. The middle graph has no such drawing. Not all vertices have the same degree.

12.15.   The statement is false. *Salvage:* There are five regular planar graphs. (These correspond, of course, to the five Platonic solids. See Challenge 13.2.)

Proof: Following the hint, we let s denote the number of sides forming the boundary of each region in the plane. By definition, $s \geq 3$. We also know that $V - E + F = 2$ for our drawing. Counting edges while we walk around the boundary of each face yields a total of sF edges. But each edge is counted twice in this process, so we have $sF = 2E$, or $F = 2E/s$. Now let d denote the degree of each vertex.

Summing the degrees over all the vertices yields dV. But again each edge is counted twice in this sum, so we have $dV = 2E$, or $V = 2E/d$.

We substitute into $V - E + F = 2$ to obtain $2E/d - E + 2E/s = 2$, or $E(2/d + 2/s - 1) = 2$. We analyze this equation to determine which values of s and d are possible. Because E is positive, we must have $2/d + 2/s > 1$. Because s and d are each at least 3, there are only a few possibilities that work:

$d = 3$, $s = 3$, which gives the tetrahedron;

$d = 3$, $s = 4$, which gives the cube;

$d = 3$, $s = 5$, which gives the dodecahedron;

$d = 4$, $s = 3$, which gives the octahedron;

$d = 5$, $s = 3$, which gives the icosahedron.

The method by which a regular polyhedron can be obtained from a regular planar graph, and vice versa, is explored in Challenge 13.2.

Stepping back

The result can be extended to any regular graph, not just planar graphs.

Proof: Let G be a regular graph with V vertices and E edges. Let d be the degree of each vertex. As we did for Challenge 12.15, sum the degrees over all the vertices to get dV. Each edge is counted twice in this sum, so we have $dV = 2E$. Therefore $d = 2E/V$. Note that we never needed a planar drawing of G.

13

Visible and invisible universes
Geometric vignettes

This module begins with a look at the Platonic solids and includes a proof that there are only five. Most of the work for this proof was done in the last two challenges of Module 12. The art gallery theorem follows—an elegant result that combines computational geometry and graph theory. The module concludes with a glimpse of higher dimensions. Though the term "dimension" is not defined rigorously, the challenges develop the n-dimensional cube and finish with some fascinating results about the boundary and the center of higher dimensional cubes. This module is well-suited for a course on proof or for a senior capstone course. If students have had analysis, the material at the end of the module could be expanded.

Note that the proof of Challenge 13.9 given here assumes familiarity with trees, which were introduced in Challenges 11.11 through 11.13. The result in Challenge 13.9 is easily understood without the idea of a tree, however. The proof itself could even be developed independently.

Solutions

13.1. There are an infinite number of regular polygons. To create a regular polygon with n sides (a *regular n-gon*), mark off n equidistant points on a circle and connect adjacent points with line segments. We can't draw them all but a few are shown in the text.

13.2. There are five regular polyhedra.

Proof: Let P be a regular polyhedron. Following the hint, we imagine projecting the edge skeleton of P onto the plane, creating a planar graph. Each vertex will have the same number of incident edges, and each region will be bounded by the same number of edges. The result is thus a regular planar graph, as defined in Module 12. By the salvage of Challenge 12.15, we know there are exactly five such graphs, and therefore there are five regular polyhedra.

Remark: The five regular polyhedra were known to the ancient Greeks and are called, of course, the Platonic solids. Students may be familiar with them, but filling out the table in the next challenge will be easier if they have the names. (See the solution to the next challenge.)

13.3.   Given a regular polyhedron, there are several counting formulas that hold. Each edge borders two faces, so if we count the edges around each face, we will count each edge twice. Thus we have

(number of faces) × (number of edges per face)/2 = number of edges.

Likewise, each edge joins two vertices, so we have

(number of vertices) × (number of edges per vertex)/2 = number of edges.

In addition, we can count the number of vertex-face pairs in two different ways to obtain

(number of faces) × (number of vertices per face) = (number of vertices) × (number of faces per vertex).

From the solution to Challenge 12.15, we have the five polyhedra as named in the table below. In each case, the name reveals the number of faces. (Well, it does if we recall that the cube is also known as the *hexahedron.*) The number of edges per face is given by the value s in the solution to Challenge 12.15.

Regular Solid	Number of Faces	Number of Edges	Number of Vertices
Tetrahedron	4	4	4
Cube	6	12	8
Octahedron	8	12	6
Dodecahedron	12	30	20
Icosahedron	20	30	12

We notice several features in this table. The cube and octahedron have the same number of edges but opposite numbers of faces and vertices. The same is true for the dodecahedron and the icosahedron. The tetrahedron stands alone in this regard, but it does have the same number of faces, edges, and vertices. This pattern of faces and vertices leads to the idea of duality, which is introduced in the next challenge.

13.4.   *Proof:* If P is a polyhedron in which every vertex has e incident edges, then every face of Dual(P) will have e sides. Because P is a regular polyhedron with all faces congruent and all angles between faces equal, the faces of Dual(P) will also be regular. In addition, if every face of P has s sides, then every vertex of Dual(P) will have s incident edges. Therefore Dual(P) is also a regular polyhedron.

Looking at the chart from Challenge 13.3, it is clear that the cube and the dodecahedron are duals of each other, as are the dodecahedron and the icosahedron. The tetrahedron is self dual.

13.5. The statement is false. *Salvage:* If *P* is a regular polyhedron, the Dual(Dual(*P*)) is a smaller version of *P*.

Proof: We know from Challenge 13.4 that Dual(*P*) is also a regular polyhedron. From the proof of Challenge 13.4, we also know that number of sides per face and number of edges per vertex in Dual(*P*) are the reverse of those in *P*. Thus in Dual(Dual(*P*)) we have switched back to exactly the parameters of *P*. The only difference is that the actual length of the edges in Dual(Dual(*P*)) has shrunk, yielding a smaller version of *P*. Some might argue that size doesn't really matter, and therefore Dual(Dual(*P*)) = *P*.

13.6.

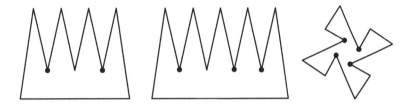

The first gallery requires two cameras, the second requires three, and the last one four. The figure shows one possible placement for the cameras in each case. If a polygon has *n* vertices, it appears that as many as *n*/3 cameras might be needed. Do we ever need more than *n*/3? Even a comb-shaped polygon with more vertices than those in the figure still requires only *n*/3 cameras. The second figure shows a gallery with 18 vertices that still only requires 6 cameras.

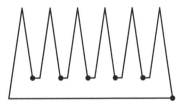

13.7. The statement is a theorem. *Extension:* The statement is true for all polygons.

Proof: (Caution: This is a very deep theorem. All we are assuming is that students will give an intuitive proof such as the one we give here.) Consider a vertex *v* adjacent to vertex *w*. Imagine a ray from *v* extending through the edge *vw*. Start sweeping the ray *vw* through the interior of the polygon. Eventually this ray must intersect another vertex. If we let *u* be the first such vertex, then the line segment *wu* must lie in the interior of *P* (as might *vu*).

13.8. The statement is false. *Counterexample:* Given a square, we can partition it in two different ways by drawing in each of the two diagonals.

Salvage: Given any polygon *P* in the plane, it is possible to partition the interior of *P* into triangles so that the vertices of *P* are precisely the vertices of the triangles.

Proof: If P is a triangle, we are done. So suppose P has at least four vertices. By Challenge 13.7, we know there is a spanning arc between two nonadjacent vertices of P. This arc divides P into two polygons, each with fewer vertices than P. If either of these polygons has four or more vertices, again, find a spanning arc between nonadjacent vertices, then repeat the process. Because the number of vertices in the subdivided polygons is reduced at each step, this process will eventually end with a complete triangulation of P.

Remark: A more formal proof requires an expanded form of mathematical induction, called *strong induction*. Strong induction works basically the same way as regular induction, except the induction hypothesis is stronger. Instead of assuming the result P holds for some $n = k$, we assume P holds for all n, $n_0 \leq n \leq k$, for some k. (The value n_0 is the smallest value of n for which P holds.) Then, as with a regular induction proof, we proceed to show the result P holds for $k + 1$. Note that while the induction hypothesis here is stronger, it really only represents what makes regular induction work in the first place. Using the domino metaphor, the kth domino falls only because all the preceding dominoes have fallen. Here is a proof of the salvage of 13.8 using strong induction.

Proof: We use strong induction on n, where n is the number of vertices of P. Our base case is $n_0 = 3$. If P has three vertices, then P is a triangle, and the result holds trivially. For our induction assumption, we assume that the result holds for any polygon with k or fewer vertices. Now we must show that the result holds for any polygon with $k + 1$ vertices. Let P be such a polygon, and let v be a vertex of P. By the extension of Challenge 13.7, there exists a spanning arc from v to some other vertex of P, call it w, that is not adjacent to v. Because v and w are not adjacent, the arc vw divides P into two polygons, each with k or fewer vertices. By our induction hypothesis, each of these two polygons can be partitioned in triangles, giving a partition of the original polygon. So by strong induction, the result holds.

13.9.   The statement is a theorem. *Extension:* Given any triangulated polygon P in the plane, there exists a way of painting each vertex one of three colors such that all three colors appear on each triangle. Moreover, there exists a way of painting each edge of P and each spanning arc within P one of three colors so that all three colors appear on each triangle.

Proof: Let P be a triangulated polygon. We will construct a tree T (see Module 11) associated with P as follows: Let each triangle in P be a vertex in T. Two vertices will be adjacent in T if they share a common edge. Clearly T is a connected graph. We see further that T has no cycles. If it did, then the corresponding triangles in P would surround a "hole" inside P, which is not possible in a polygon. Thus T is indeed a tree. Note that if P has only three vertices, then T consists of a single vertex.

To color the vertices of P, pick a vertex v in T and color the vertices of the corresponding triangle in P red, blue, and green. Now proceed to any adjacent vertex, u. The corresponding triangle has an edge in common with v's triangle and so already has two vertices colored. Color the remaining vertex the third color. Continue this process, moving from fully colored triangles to adjacent triangles with uncolored vertices. Because T is connected, we will reach all triangles. Because T has no cycles, we will never produce a coloring conflict.

To color the edges and arcs of P, we follow an analogous algorithm. The result follows.

13.10. The statement is a theorem (the Art Gallery Theorem). *Extension:* There is no natural extension to this theorem, though there are many variations. Students might want to explore what happens if the polygon has only right angles (both concave and convex).

Proof: First triangulate P. Notice that every point in the interior of P lies in a triangle or on the edge of a triangle. Note also that for every triangle, each point in its interior and on its edges is visible from each vertex. By Challenge 13.9, we can paint each vertex with one of three colors, say red, green, and blue, so that all three colors appear on each triangle. Let x_r, x_g, x_b denote the number of vertices painted red, green, and blue, respectively. Then $x_r + x_g + x_b = n$, so at least one of the x's is less than or equal to $n/3$, say, x_r. Thus $x_r \leq [n/3]$. Every triangle has a red vertex, so every point in the interior of P is visible from a red vertex and the result holds.

13.11. $\mathbb{R}^N$ is the set of all N-tuples of real numbers: $(x_1, x_2, \ldots, x_N)$. To locate a point in $\mathbb{R}^N$, we clearly need exactly the N numbers: $x_1, x_2, \ldots, x_N$. So by our informal definition, $\mathbb{R}^N$ has dimension N.

13.12. To create a 3-dimensional cube, we would ink up an entire 2-dimensional solid cube (square) and drag it 1 unit in a new perpendicular dimension. Note that to do this, we must be in a 3-dimensional space. To create a 4-dimensional cube, we ink up an entire 3-dimensional solid cube and drag it 1 unit in a new, perpendicular direction. Note that to do this, we need to be in a 4-dimensional space. In theory, we can continue this process as many times as we wish, creating n-dimensional cubes in n-dimensional space.

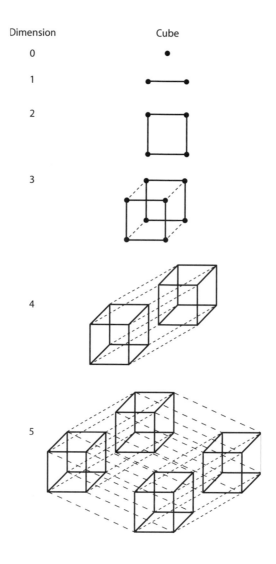

13.13.

Dimension n	Number of vertices	Number of edges	Number of 2-D "faces"	Number of 3-D "faces"	Number of 4-D "faces"
0	1	0	0	0	0
1	2	1	0	0	0
2	4	4	1	0	0
3	8	12	6	1	0
4	16	32	24	8	1
5	32	80	80	40	10
⋮	⋮	⋮	⋮	⋮	⋮
n	2^n	See below.	…	…	…

$C(n)$ is created by "dragging" a copy of $C(n-1)$ in a new, perpendicular dimension, which doubles the number of vertices. Thus it is easy to derive the formula 2^n for the number of vertices in $C(n)$. The remaining values of the nth row of the table can be expressed recursively in terms of the values in the previous row. As with the vertices, the number of edges in $C(n)$ is double the number in $C(n-1)$, but in addition, each vertex of $C(n-1)$ will drag out an extra new edge of $C(n)$. Therefore the number of edges in $C(n)$ equals twice the number of edges of $C(n-1)$ plus the number of vertices in $C(n-1)$. Similarly, the number of 2D faces in $C(n)$ is double the number in $C(n-1)$ plus each edge of $C(n-1)$ will drag out an additional new 2D face of $C(n)$. Therefore the number of 2D faces in $C(n)$ equals twice the number of 2D faces of $C(n-1)$ plus the number of edges in $C(n-1)$. This same recursive pattern holds for the remaining entries in the nth row. Thus we have that for $n \geq 2$, an entry in the table equals the sum of twice the entry directly above it plus the entry above and to the left.

Note: It is possible to solve this recurrence relation, though we won't do it here. The number of k-faces in an n-dimensional cube is $\binom{n}{k} 2^{n-k}$.

13.14. *Proof:* We will extend the idea of volume to n dimensions in the natural way: An n-dimensional cube with side length a will have volume a^n. Thus $C(n)$ has volume $1^n = 1$, and a cube with side length $1/2$ will have volume $1/2^n$. Thus the proportion of points in $C(n)$ that lie in the inner, concentric cube with side length $1/2$ is exactly $(1/2^n)/1 = 1/2^n$. As n increases, this proportion goes to 0, so the probability that a randomly selected point lies outside the inner cube and thus, by definition, near the boundary, approaches 1.

13.15. *Proof:* We have $C(2m)$ and $B(2m, 1/2)$, both centered at the origin. The cube has side length 1 and the ball has radius $1/2$, so the surface of the ball will be tangent to the faces of the cube. Thus the ball is inscribed in the cube. The volume of $C(n)$ is always $1^n = 1$, so $C(2m)$ does not change as $m \to \infty$. Now consider the volume of $B(2m,1/2)$. We have

$$V(B(2m,1/2)) = \frac{\pi^m (1/2)^{2m}}{m!} = \left(\frac{\pi}{4}\right)^m \cdot \frac{1}{m!}.$$

Because $\frac{\pi}{4} < 1$, both factors above go to 0 as $m \to \infty$, so $V(B(2m, r))$ will go to 0. Thus as $m \to \infty$, the ball shrinks to a space with zero volume, implying that the central space inside the cube gets very small indeed!

Stepping back

If C is a circle of radius r, then the area of C equals the integral of the circumference, as follows:

$$\int_0^{2\pi} 2\pi r \, dr = \pi r^2.$$

This makes sense if we think of the integral as the limit of a Riemann sum. Let Δr be the width of an annulus of radius approximately r. If we cut the annulus and stretch it out straight, it is close to a rectangle of length $2\pi r$ and thickness Δr. Thus its area is approximately $2\pi r\, \Delta r$. If we partition C into concentric annuluses, the area of C will equal the sum of the areas of the annuluses, each of which we approximate with $2\pi r\, \Delta r$. When we take a limit as $\Delta r \to 0$, we get the integral formula above.

In general, when we integrate the formula for the surface area of a ball in n dimensions, we will get the volume of that ball. (Try this in the case $n = 3$.) Conversely, differentiating the formula for volume gives the formula for surface area.

14

A synergy between geometry and numbers

Circles and Pythagorean triples

This module begins with an elegant proof of the Pythagorean Theorem, followed by a generalization that leads to the Law of Cosines. Now that they can discover proofs on their own, students will enjoy seeing the familiar results of Challenges 14.3 through 14.6 with greater understanding and confidence than they might have experienced in high school. Challenges 14.7 through 14.10 develop the subtle relationship between Pythagorean triples and points on the unit circle with rational coordinates, a relationship that underlies the field of arithmetic geometry. This material is particularly appealing for a senior capstone course or a course on mathematics for teachers.

Solutions

14.1. Let the triangle side lengths be a, b, and c, as labeled in the figure. We know the small square fits perfectly in the middle, because its sides have length $b - a$ and the triangles are right triangles. We also know the larger square is indeed a square, because the acute angles of the triangle sum to 90 degrees.

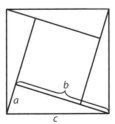

14.2.   Because the small square has side length $b - a$, the figure comprises a square with side length a and a square of side length b.

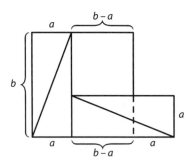

14.3.   *Proof of the Pythagorean Theorem:* Consider an arbitrary right triangle with leg lengths a and b and hypotenuse c. Construct a figure as in Challenge 14.1 and note that it has area c^2. The same five shapes can be rearranged as in Challenge 14.2 to form a figure with area $a^2 + b^2$. Because the two figures were formed with exactly the same five shapes, with no overlap and no spaces, the two figures must have the same area. Thus we have $a^2 + b^2 = c^2$ for any right triangle.

14.4.   *Proof:* In the figure, we have two right triangles, so the Pythagorean Theorem implies that $a'^2 + h^2 = b^2$ and $(a - a')^2 + h^2 = c^2$. Rearranging the terms, we see that $h^2 = b^2 - (a')^2$ and $h^2 = c^2 - (a - a')^2$. Thus $b^2 - (a')^2 = c^2 - (a^2 - 2aa' + (a')^2)$, which is equivalent to $a^2 + b^2 = c^2 + 2aa'$.

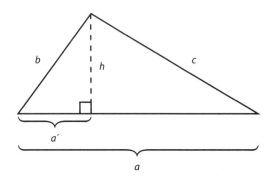

14.5.   The statement is false. *Counterexample:* Consider a triangle with two sides of length $a = b = 1$ meeting at $\theta = 60°$ ($\pi/3$ radians). Then $\cos\theta = 1/2$. So as written, the statement would imply that the third side of the triangle had length $c = 1^2 + 1^2 + 2 \cdot 1 \cdot 1 \cdot (1/2) = 3$, which is impossible. *Salvage:* If the sides of a triangle have lengths a, b, and c and if θ is the angle between the sides having lengths a and b, then $c^2 = a^2 + b^2 - 2ab\cos\theta$.

Proof: Rearranging the result of Challenge 14.4, we obtain $c^2 = a^2 + b^2 - 2aa'$. Looking at the figure in Challenge 14.4, we also note that $a' = b\cos\theta$. Thus, by substitution, we see that $c^2 = a^2 + b^2 - 2ab\cos\theta$.

14.6.   The converse of the Pythagorean Theorem is true: A triangle with side lengths $0 < a \leq b < c$ is a right triangle if and only if $a^2 + b^2 = c^2$.

Proof: Consider a triangle T with side lengths $0 < a \leq b < c$. If T is a right triangle, we already know that $c^2 = a^2 + b^2$. So suppose T satisfies the property that $a^2 + b^2 = c^2$. By the salvage of Challenge 14.5, we also have $c^2 = a^2 + b^2 - 2ab \cos \theta$. Thus we know that $2ab \cos \theta = 0$. We know that a and b are nonzero, so we must have $\cos \theta = 0$. Thus $\theta = \frac{\pi}{2}$ radians. Therefore T is a right triangle, and the converse of the Pythagorean Theorem is true.

14.7.   The statement is false. *Counterexample:* Consider the Pythagorean triples $(3, 4, 5)$ and $(6, 8, 10)$. Then $f(3, 4, 5) = (3/5, 4/5)$ and $f(6, 8, 10) = (3/5, 4/5)$. Thus, the map is not one-to-one. Note furthermore that the function f is not even defined for the trivial Pythagorean triple $(0, 0, 0)$. *Salvage:* If we let the domain of the function f be nontrivial Pythagorean triples with $z > 0$ and x, y, and z relatively prime, then $f(x, y, z) = (x/z, y/z)$ is a one-to-one and onto map to rational points on the unit circle.

Proof: First we observe that f does indeed map Pythagorean triples to rational points on the unit circle, as follows: If (x, y, z) is a Pythagorean triple with $z \neq 0$, then $x^2 + y^2 = z^2$, and we can write $x^2/z^2 + y^2/z^2 = 1$. Thus if we let $f(x, y, z) = (x/z, y/z) = (X, Y)$, we have $X^2 + Y^2 = 1$.

   To show f is one-to-one, suppose (x_1, y_1, z_1) and (x_2, y_2, z_2) are relatively prime Pythagorean triples with $z_i \neq 0$, $i = 1, 2$, such that $f(x_1, y_1, z_1) = f(x_2, y_2, z_2)$. Then $(x_1/z_1, y_1/z_1) = (x_2/z_2, y_2/z_2)$, so we have $x_1/z_1 = x_2/z_2$ and $y_1/z_1 = y_2/z_2$. This implies, in particular, that $x_1 = z_1 \frac{x_2}{z_2}$. Because our triples are relatively prime, we know x_2 shares no common factors with z_2. So the only way that $z_1 \frac{x_2}{z_2}$ can equal the integer x_1 is for z_2 to divide z_1. Thus $z_1 = mz_2$ for some integer m, which implies $x_1 = z_1 \frac{x_2}{z_2} = mz_2 \frac{x_2}{z_2} = mx_2$. Similarly we have $y_1 = z_1 \frac{y_2}{z_2} = mz_2 \frac{y_2}{z_2} = my_2$. We also know m is positive because z_1 and z_2 are positive by hypothesis. But then m is a factor common to each of x_1, y_1, and z_1, which implies $m = 1$. Therefore $(x_1, y_1, z_1) = (x_2, y_2, z_2)$ and f is one-to-one.

   To show f is onto, suppose (X, Y) is a rational point on the unit circle. Then $(X, Y) = (a/c, b/d)$ for integers a, b, c, and d such that $(a/c)^2 + (b/d)^2 = 1$. This implies $a^2 d^2 + b^2 c^2 = c^2 d^2$, and thus (ad, bc, cd) is a Pythagorean triple. Without loss of generality, we may assume $c > 0$ and $d > 0$. We will show that reducing a/c and b/d to lowest terms will lead us to a relatively prime Pythagorean triple that also maps to (X, Y) under f. Let g be the $\gcd(a, c)$ and h be the $\gcd(b, d)$, so that $a = a'g$, $c = c'g$, $b = b'h$, and $d = d'h$. Note that we may assume g and h are positive and, therefore, that c' and d' are positive. Note also that a' and c' are relatively prime, as are b' and d'. We know that $a^2 d^2 + b^2 c^2 = c^2 d^2$. Therefore we have $(a'g)^2 (d'h)^2 + (b'h)^2 (c'g)^2 = (c'g)^2 (d'h)^2$. Dividing by $g^2 h^2$ yields $(a'd')^2 + (b'c')^2 = (c'd')^2$. Thus we have another Pythagorean triple, $(a'd', b'c', c'd')$, that f maps to (X, Y) but that is not relatively prime (see below). We will now show that, in fact, $c' = d'$, which will allow us to reduce the equation further to reach our goal.

   Rewriting $(a'd')^2 + (b'c')^2 = (c'd')^2$ as $a'^2 d'^2 + b'^2 c'^2 = c'^2 d'^2$, we see that c'^2 must divide all three terms of the equation, as must d'^2. Thus c'^2 divides $a'^2 d'^2$. Because a' and c' are relatively prime, we must have $c'^2|d'^2$. Similarly we must have $d'^2|c'^2$. Therefore $c'^2 = d'^2$, and we have $c' = d'$ because both numbers are positive. We now divide the equation $a'^2d'^2 + b'^2c'^2 = c'^2d'^2$ by d'^2 to obtain $a'^2 + b'^2 = c'^2$. Therefore (a', b', c') is a Pythagorean triple. Furthermore, it is relatively prime because a' and c' are relatively prime, as are b' and $c' = d'$. We also know that $c' > 0$ and $f(a', b', c') = (X, Y)$. Therefore f maps the set of relatively prime Pythagorean triples with $z > 0$ onto the set of rational points on the unit circle.

Remark: The restricted set of Pythagorean triples described in the salvage sounds very much like primitive triples, but note that we allow nonpositive values of x and y, whereas primitive triples are all positive. Note further that triples of the form $(1, 0, 1)$ are relatively prime because 1 is the greatest common divisor of any two of the numbers. We can restrict the domain of f to primitive Pythagorean triples to obtain a one-to-one, onto map to those rational points on the unit circle, other than $(1, 0)$ and $(0, 1)$, that lie in the first quadrant. We can easily revise the proof above to prove this result, which we will use in Challenge 14.10.

14.8. The statement is a theorem. *Extension:* The result applies not just to the unit circle but also to any conic section whose graph contains at least two rational points. We will not prove this result, though the last challenge of the module (*Stepping back*) asks a related question for a special class of ellipses.

Proof of original statement: Let (x, y) be a rational solution to $X^2 + Y^2 = 1$ other than $(-1, 0)$. Then the line through $(-1, 0)$ and (x, y) has slope $\frac{y}{x+1}$, which is rational because x and y are rational. For the converse, let m be rational. Then the line through $(-1, 0)$ with slope m has equation $y = m(x + 1)$. To find where this line intersects the unit circle, we substitute to obtain $x^2 + m^2(x + 1)^2 = 1$, which becomes $(m^2 + 1)x^2 + 2m^2x + m^2 - 1 = 0$. Using the quadratic formula, we obtain

$$x = \frac{-2m^2 \pm \sqrt{4m^4 - 4(m^2 + 1)(m^2 - 1)}}{2(m^2 + 1)} = \frac{-2m^2 \pm 2}{2(m^2 + 1)}.$$

In one case, we have $x = -1$, giving us the point $(-1, 0)$. In the other case, we have $x = \frac{1-m^2}{1+m^2}$, which yields $y = m(\frac{1-m^2}{1+m^2} + 1) = \frac{2m}{1+m^2}$. These values are rational because m is rational, and the result holds.

14.9. The statement is false. *Counterexample:* Let $m = 1/1 = 1$. The line with slope 1 passing through $(-1, 0)$ also intersects the unit circle at $(0, 1)$, not at $(0, 1/2)$ as the statement claims. We need to insert a factor of 2 appropriately. *Salvage:* Suppose that $m = s/r$. Then the line of slope m passing through $(-1, 0)$ also intersects the unit circle at

$$(X, Y) = \left(\frac{r^2 - s^2}{r^2 + s^2}, \frac{2rs}{r^2 + s^2} \right).$$

Proof: From the proof of Challenge 14.8, we know that the line of slope m passing through $(-1, 0)$ also intersects the unit circle at

$$(X, Y) = \left(\frac{1 - m^2}{1 + m^2}, \frac{2m}{1 + m^2} \right).$$

Letting $m = s/r$, we obtain

$$X = \frac{1 - m^2}{1 + m^2} = \frac{1 - (s/r)^2}{1 + (s/r)^2} \quad \text{and} \quad Y = \frac{2m}{1 + m^2} = \frac{2s/r}{1 + (s/r)^2}.$$

Multiplying both fractions by r^2/r^2, we obtain

$$X = \frac{r^2 - s^2}{r^2 + s^2} \quad \text{and} \quad Y = \frac{2rs}{r^2 + s^2}.$$

14.10. *Proof:* Challenge 14.7 implies that an infinite number of rational points in the first quadrant on the unit circle will give rise to an infinite number of primitive Pythagorean triples. Challenge 14.8 tells us that every line through $(-1, 0)$ with rational slope will intersect the unit circle in a second rational point. Clearly two lines through $(-1, 0)$ with different slopes will have different points of intersection elsewhere on the unit circle. Therefore, because there are infinitely many positive rational numbers, and thus infinitely many lines through $(-1, 0)$ with different slopes, we have infinitely many primitive Pythagorean triples.

The salvage of Challenge 14.9 tells us that for any rational number $m = s/r$, the point

$$(X, Y) = \left(\frac{r^2 - s^2}{r^2 + s^2}, \frac{2rs}{r^2 + s^2} \right)$$

is a rational point on the unit circle. So $(x, y, z) = (r^2 - s^2, 2rs, r^2 + s^2)$ is a Pythagorean triple. Because we have $0 < s < r$, we know x, y, and z are all positive. To show our triple is primitive, we just need to show that x, y, and z are relatively prime. We know that r and s are relatively prime and that one of them is even; so the other is odd. Therefore one of r^2 and s^2 is even and the other is odd, which implies that both $r^2 - s^2$ and $r^2 + s^2$ are odd. Now suppose p is a prime dividing both $r^2 - s^2$ and $r^2 + s^2$. Then p divides the sum of these two quantities. So we have $p \,|\, 2r^2$. We know p must be odd, therefore $p \,|\, r^2$. Because p is prime, we must have $p \,|\, r$. But p also divides the difference $r^2 + s^2 - (r^2 - s^2) = 2s^2$; so by similar reasoning $p \,|\, s$, which violates r and s being relatively prime. Therefore we have $r^2 - s^2$ and $r^2 + s^2$ relatively prime.

To show that $2rs$ is also relatively prime to $r^2 - s^2$ and $r^2 + s^2$, we note once again that 2 does not divide $r^2 - s^2$ or $r^2 + s^2$. Now suppose $p \,|\, 2rs$ and $p \,|\, r^2 - s^2$ for some odd prime p. Then because p is prime, we must have $p \,|\, r$ or $p \,|\, s$. If $p \,|\, r$, then $p \,|\, r^2$. Then because $p \,|\, r^2 - s^2$, we must also have $p \,|\, s^2$, which implies $p \,|\, s$. This violates r and s being relatively prime; therefore $2rs$ and $r^2 - s^2$ must also be relatively prime. A similar argument shows $2rs$ and $r^2 + s^2$ are relatively prime. Therefore the Pythagorean triple $(r^2 - s^2, 2rs, r^2 + s^2)$ is primitive.

Stepping back

This challenge generalizes the result of Challenge 14.8 to a special class of ellipses.

Call the equation $x^2 + py^2 = z^2$ equation P. We first note that $(0, 0, 0)$ is a trivial solution to equation P. For any nontrivial integer solution (x, y, z), we must have $z \neq 0$, and so $\frac{x^2}{z^2} + p\frac{y^2}{z^2} = 1$. Therefore $\left(\frac{x}{z}, \frac{y}{z}\right)$ is a rational solution to $X^2 + pY^2 = 1$, an ellipse. We will follow the methods of Challenges 14.7 and 14.8 and use rational points on this ellipse to find solutions to equation P.

The point $(-1, 0)$ lies on the ellipse and corresponds to the solution $(-1, 0, 1)$ to equation P. Consider the line $y = m(x + 1)$ with slope m through the point $(-1, 0)$. For each real number slope m, this line intersects the ellipse in exactly one other point. Conversely every point on the ellipse not equal to $(-1, 0)$ determines a unique line through $(-1, 0)$ with slope m. Thus there is a one-to-one correspondence between points on the ellipse and slopes m of the line $y = m(x + 1)$.

Now suppose the slope m is rational. Let (X, Y) be a point where the line $y = m(x + 1)$ intersects the ellipse. We will show that (X, Y) is a rational point. We have $Y = m(X + 1)$. So we substitute to obtain $X^2 + p(m(X + 1))^2 = 1$, which expands to $(pm^2 + 1)X^2 + 2pm^2 X + pm^2 - 1 = 0$. Thus we have

$$X = \frac{-2pm^2 \pm \sqrt{(2pm^2)^2 - 4(pm^2 + 1)(pm^2 - 1)}}{2(pm^2 + 1)} = \frac{-pm^2 \pm 1}{pm^2 + 1}.$$

Therefore $X = -1$ (and thus $Y = 0$, as we already knew), or $X = \frac{-pm^2 + 1}{pm^2 + 1} = \frac{1 - pm^2}{1 + pm^2}$, which is clearly rational because m is rational. The corresponding value of Y for the second value of X is

$$Y = m\left(\frac{1 - pm^2}{1 + pm^2} + 1\right) = m\left(\frac{1 - pm^2 + 1 + pm^2}{1 + pm^2}\right) = \frac{2m}{1 + pm^2},$$

which is also clearly rational. Thus we have rational slopes yielding rational points on the ellipse.

Conversely if $(X, Y) \neq (-1, 0)$ is a rational point on the ellipse, then the line determined by these two points has slope $Y/(X + 1)$, which is rational because X and Y are rational. Therefore rational slopes determine rational points on the ellipse and vice versa.

Thus every solution to equation P corresponds to a rational point on the ellipse and, therefore, corresponds to a rational slope. Conversely every rational slope gives rise to, as we shall see, infinitely many solutions to equation P. Let slope $m = s/r$ for integers s and r, with $r \neq 0$. We know the line $y = m(x + 1)$ through $(-1, 0)$ intersects the ellipse at the rational point

$$(X, Y) = \left(\frac{1 - pm^2}{1 + pm^2}, \frac{2m}{1 + pm^2}\right) = \left(\frac{1 - p\left(\frac{s}{r}\right)^2}{1 + p\left(\frac{s}{r}\right)^2}, \frac{2\left(\frac{s}{r}\right)}{1 + p\left(\frac{s}{r}\right)^2}\right) = \left(\frac{r^2 - ps^2}{r^2 + ps^2}, \frac{2rs}{r^2 + ps^2}\right).$$

This point gives rise to the solution $(r^2 - ps^2, 2rs, r^2 + ps^2)$ to equation P. Note that for every distinct pair of integers s and r for which the quotient $s/r = m$, we get a different solution to equation P. Thus each slope m gives rise to infinitely many solutions. Furthermore any solution corresponds to a rational point and must therefore be expressible in the form $(r^2 - ps^2, 2rs, r^2 + ps^2)$. Therefore any choice of integers r and s yields a solution to equation P. We even allow $r = 0$, which corresponds to the vertical line through $(-1, 0)$ and gives rise to solutions with $y = 0$.

Thus we conclude that all solutions to $x^2 + py^2 = z^2$ have the form $(r^2 - ps^2, 2rs, r^2 + ps^2)$, where r and s are integers.

15

The mathematical mysteries within a sheet of paper
Unfolding pattern and structure

By studying patterns in a folded piece of paper, students explore mathematical structure using only simple observations and careful reasoning. This is a stand-alone module that works well anywhere in a course. It has even been used as the first module in a discrete math course. The topic will be new to most students, putting them all on the same footing. The paper-folding process is very simple, allowing students to experiment easily. But the resulting structures quickly get complicated, so students will be challenged. Many of the challenges can be reasoned out with careful thinking, rather than formal techniques that students may have not yet studied.

The *Stepping back* challenge assumes some familiarity with Turing machines. Depending on students' backgrounds, you may wish to give them guidance after reading the solution presented here. In addition, Challenges 15.9 and 15.10 involve limits and functions. Experience with these topics in a calculus course should be sufficient for students who have not studied Module 6 or Module 16.

Solutions

15.1. *Proof:* For a sheet of paper folded in this manner, a *strip* is a segment of the paper that spans from one fold to the next or from the edge of the paper to a fold. Viewed edgewise, a strip spans the width from left to right. On either end of a strip, it terminates in a fold or an endpoint of the paper. When the strip doesn't form an endpoint on the right (left), it forms a right (left) fold. The next strip starts there and crosses the width to form a fold of opposite parity or an endpoint. Thus the folds alternate between right and left when viewed edgewise. Because the bottom endpoint is on the left, the alternating sequence of folds begins R, L, R, ...

15.2. *Proof:* We use induction. In the base case, we have a sheet with 0 folds and 1 strip. Assume that at the nth stage, the sheet has 2^n strips. The $(n+1)$th stage effectively folds each strip from the nth stage in half, creating two strips from every strip at the nth stage, for a total of 2^{n+1} strips. The result follows.

15.3. The statement is a theorem. *Extension:* The result does not depend on left or right folding. After a sequence of n folds in a sheet of paper (left or right folds), the total number of folds equals $2^{n-1} + 2^{n-2} + \cdots + 2^1 + 2^0$.

Proof of extension: The number of folds in the edge introduced by a fold of the sheet is equal to the number of strips in the sheet before the additional fold. Thus, the first fold of the paper introduces $2^0 = 1$ folds in the edge. The second fold introduces $2^1 = 2$ folds in the edge, and at that time, there are $2^1 + 2^0 = 3$ folds in the edge. This pattern continues, so the number of folds in the paper after n folds, right or left, equals $2^{n-1} + 2^{n-2} + \cdots + 2^1 + 2^0$. This argument can be formalized using induction.

15.4. *Proof:* Consider the proof of the extension of 15.3. The most recent fold creates 2^{n-1} right folds, and all the earlier right folds become left folds. The number of left folds equals the total number of folds minus the number of right folds—that is, $2^{n-2} + 2^{n-3} + \cdots + 2^1 + 2^0$. This is the sum of a finite geometric series, which simplifies to $\frac{2^{n-1}-1}{2-1} = 2^{n-1} - 1$.

15.5. *Proof:* Look closely at a right fold. It divides a strip into two halves that, when the sheet is unfolded, become a left half and a right half of either a valley or a ridge. Viewing the folded sheet on edge, if the left half is on the bottom and the right half on the top, then in the Life sequence, the shape of the fold is a valley ($\vee$.) If the left half is on the top, then the shape of the fold is a ridge ($\wedge$). (Try this with a sheet of paper and see for yourself.)

By Challenge 15.2, we know that new folds are added in the middle of existing strips. The existing strips are folded so that, starting with the strip on the bottom, the new folds have left side on bottom and right side on top, then left side on top and right side on bottom, and so on. Thus an additional right fold on the sheet creates new folds of alternating orientation when viewed on edge. When the sheet is unfolded, these new folds have shape $\vee, \wedge, \vee, \wedge, \ldots$.

15.6. *Proof:* By Challenge 15.5, we know that the sequence of new folds is $\vee, \wedge, \vee, \ldots$, and by Challenge 15.2, we know that new folds are added in the middle of existing strips between existing folds. Thus each new right fold adds the sequence $\vee, \wedge, \vee \ldots$, perfectly shuffled with the existing folds.

15.7. The first $2^{n-1} - 1$ folds of the nth stage of the Life sequence are identical to the folds of the $(n-1)$th stage sequence.

Proof:

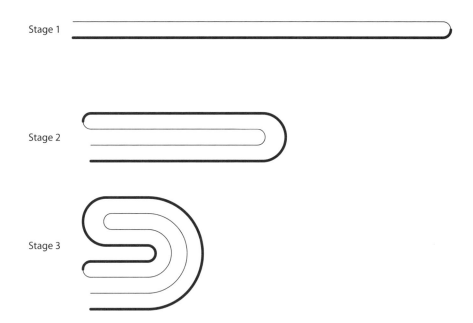

Stage 1

Stage 2

Stage 3

This result is easy to see if we look at the early stages of the right-folding sequence. Stage 1, the left half of the paper is on the bottom (in bold). Treat it as a piece of paper that has not yet been folded, so it is in Stage 0. As subsequent folds are made in Stages 2 and 3, the left half of the original paper is always being folded in exactly a right-folding sequence but a stage behind the original paper. The result follows.

15.8.   The last $2^{n-1} - 1$ folds of the nth stage of the Life sequence are identical to the folds of the $(n-1)$th stage sequence but in reverse order and inverted.

Proof: We look at the same figure as for Challenge 15.7. This time, look at the right half of the original paper. At Stage 1, the right half (nonbold) is a piece of paper that has not been folded, but it is upside down on top of the left half. So its right and left endpoints are inverted from their final positions in the original paper. As subsequent folds are made, the right half of the original paper is being folded in a right-folding sequence but a stage behind the original paper and upside down. When the original paper is unfolded at the nth stage, the right half is flipped over, reversing the order of folds and inverting valleys to ridges and ridges to valleys.

15.9.   *Proof:* To get a sense of what the paper-folding sequence looks like, here are the sequences PFS_n for $n = 1, 2, 3, 4,$ and 5.

1

1 1 0

1 1 0 1 1 0 0

1 1 0 1 1 0 0 1 1 1 0 0 1 0 0

1 1 0 1 1 0 0 1 1 1 0 0 1 0 0 1 1 1 0 1 1 0 0 0 1 1 0 0 1 0 0

By Challenge 15.7, the first $2^n - 1$ folds of the nth stage are the same as the sequence of folds in the $(n-1)$th stage. Thus the digits of PFS_{n-1} are repeated as the digits in the first half of PFS_n. Furthermore, these digits will be repeated at the beginning of all PFS_k for $k \geq n$. Thus given $m \geq 1$, for all n such that $m \leq 2^{n-1} - 1$, we have $PFS_n(m)$ in the first half of PFS_n. Therefore $PFS_n(m)$ will remain unchanged for all these n, which implies that $\lim_{n \to \infty} PFS_n(m)$ exists. Because this limit exists, the value of f_m makes sense.

15.10.   *Proof:* The function $\frac{x}{1-x^4}$ has power series expansion $x \cdot \sum_{m=0}^{\infty} x^{4m} = \sum_{m=0}^{\infty} x^{4m+1}$. We will compare the coefficients of this power series with those of the power series for $\mathcal{F}(x) - \mathcal{F}(x^2)$.

Given that $\mathcal{F}(x) = \sum_{m=1}^{\infty} f_m x^m$, we have $\mathcal{F}(x^2) = \sum_{m=1}^{\infty} f_m x^{2m}$.

Look at just the coefficients of the two sequences listed as strings of 0's and 1's:

$$\mathcal{F}(x): \ 110110011100100111011000110010 0\ldots$$
$$\mathcal{F}(x^2): \ 010100010100000101010000010000 0\ldots\ .$$

Taking the difference, we find the coefficients so far to be

$$\mathcal{F}(x) - \mathcal{F}(x^2): \ 100010001000100010001000 1000.$$

These are the coefficients of the power series $x \sum_{m=0}^{\infty} x^{4m}$. Therefore assuming the pattern continues, we conclude that $\mathcal{F}(x) - \mathcal{F}(x^2) = \sum_{m=0}^{\infty} x^{4m+1}$.

Stepping back

We begin with a ticker tape stretching to the right and consisting of infinitely many cells. Each cell can hold a single digit. When our Turing machine starts, the first cell contains the digit 1 and the remaining cells are blank.

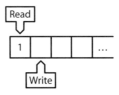

Starting state of Turing machine

Our machine has a Read head that scans one cell at a time, moving from left to right. The Write head is positioned at the first blank cell. The Write head will insert one digit per cell, according to the following instructions:

If the Read head scans...	the Write head inserts...
1	3, then 2
2	4, then 2
3	3, then 1
4	4, then 1

The next figure shows the state of the Turing machine at the end of each of the first three steps.

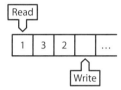

State of Turing machine after Step 1,

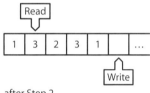

after Step 2,

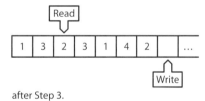

after Step 3.

The Turing machine never stops, thus creating an infinite string of digits: 13231423132414231323 14241…. . This string is converted to a string of 0's and 1's by reducing each digit mod 2: 11011001110 01001110110001…. . Notice that the digits in this string match the corresponding digits of the paper-folding sequence. In general, it turns out that the string of 0's and 1's in the first $2^n + 1$ cells matches PFS_n precisely for each n. We omit the formal proof.

16

Take it to the limit
An initial approach to analysis

This module offers students a teasing glimpse into the world of analysis. Challenges 16.1 through 16.7 introduce limits of sequences using the formal ε definition. Experience with proofs using this definition is an important part of any student's preparation for a course in real analysis. The module continues with a careful look at the Cantor set, leading to some very interesting challenges. The *Stepping back* question is challenging and could be omitted, though ambitious students may enjoy tackling it.

We choose to be more casual about some technicalities in the solutions for this module, treating students with more sophistication. For example, the proofs for Challenges 16.8 and 16.9 use induction. By this time, students should be very familiar with how induction works, so we don't need to spell out every detail of an induction proof. It's a great moment when you tell students that they have the mathematical maturity and credibility to leave out more details from their proofs. Be sure, however, to insist that they include important details pertaining to newly introduced ideas, such as the definition of a converging sequence. We also note that Challenge 16.14 calls upon the result of Challenge 7.7.

The main point of this module is to foreshadow the style of real analysis. If you wish, embrace the informalism and use the broad ideas introduced here as a springboard for discussion about the nature of analysis. With all its engaging challenges about the Cantor set, this module would be an excellent component for a senior capstone course.

Solutions

16.1. The limit exists and equals 2/3 (see the *Remark* following the proof).

Proof: To show that the limit equals 2/3, we need to show that given any $\varepsilon > 0$, there exists an integer N such that for all $n \geq N$, it follows that

$$\left| \frac{2n+1}{3n-500} - \frac{2}{3} \right| < \varepsilon.$$

To find N as a function of ε, it is often easiest to work backward from what we want. In this case, we want

$$\left| \frac{2n+1}{3n-500} - \frac{2}{3} \right| = \left| \frac{3(2n+1)}{9n-1500} - \frac{2(3n-500)}{9n-1500} \right| = \left| \frac{1003}{9n-1500} \right| < \varepsilon.$$

We solve this inequality for n in terms of ε and deduce that

$$-\varepsilon < \frac{1003}{9n-1500} < \varepsilon$$

$$-9n\varepsilon + 1500\varepsilon < 1003 < 9n\varepsilon - 1500\varepsilon$$

$$0 < 9n\varepsilon - 1500\varepsilon + 1003 < 18n\varepsilon - 3000\varepsilon$$

$$3000\varepsilon + 1003 - 1500\varepsilon < 9n\varepsilon$$

$$\frac{1003+1500\varepsilon}{9\varepsilon} < n.$$

By choosing N to be an integer greater than $\frac{1003+1500\varepsilon}{9\varepsilon}$, we know that whenever $n \geq N$, we will have $|a_n - 2/3| < \varepsilon$. Therefore $\lim_{n \to \infty} a_n = 2/3$.

Remark: Many students may be comfortable computing this limit using reasoning similar to the following:

$$\lim_{n \to \infty} a_n = \lim_{n \to \infty} \frac{2n+1}{3n-500} = \lim_{n \to \infty} \frac{2+1/n}{3-500/n} = 2/3,$$

because both $1/n$ and $500/n$ go to 0 as $n \to \infty$. Reassure them that using such an approach to find the value of the limit is fine at this stage if they then use the definition of a limit to prove their answer is valid, as the challenge instructs. Point out that whatever method they use to find a value for the limit probably depends on basic results about limits that have not yet been established using the rigorous definition given here.

16.2.   The limit does not exist.

Proof: We observe that the numerator of a_n can be factored into $(n-1)(n+1)$, giving us $a_n = n - 1$. As $n \to \infty$, clearly the values of a_n increase without bound. Therefore there is no number L to which values of a_n get arbitrarily close as $n \to \infty$.

More formally, we can show that given any L, any $\varepsilon > 0$, and integer N, there exists an integer $n \geq N$ such that $|a_n - L| > \varepsilon$. Because $a_n = n - 1$, we note that $|a_n - L| = |n - 1 - L| = |n - (L+1)|$. This quantity will be greater than ε for every n greater than $L + 1 + \varepsilon$ and, thus, for any such n that is also greater than some fixed N. Therefore no L exists that satisfies the definition of a limit, and the limit does not exist.

16.3.   The statement is false. *Counterexample:* Let $a_n = n$ and $b_n = -n$. Both $\{a_n\}$ and $\{b_n\}$ diverge, but $\{a_n + b_n\}$ converges because $\lim_{n\to\infty}(a_n + b_n) = 0$. *Salvage:* Suppose $\{a_n\}$ and $\{b_n\}$ are infinite sequences. If $\{a_n\}$ converges to A and $\{b_n\}$ converges to B, then the sequence $\{a_n + b_n\}$ converges to $A + B$. If one of $\{a_n\}$ or $\{b_n\}$ converges and the other diverges, then the sequence $\{a_n + b_n\}$ diverges.

Proof: Suppose that $\{a_n\}$ converges to A and $\{b_n\}$ converges to B. Let $\varepsilon > 0$; thus $\varepsilon/2 > 0$. By definition, there are integers K and M such that for $k \geq K$, we have $|a_k - A| < \varepsilon/2$, and for $m \geq M$, we have $|b_m - B| < \varepsilon/2$. Let N be the larger of K and M. Then for any $n \geq N$, by the triangle inequality, we have

$$|(a_n + b_n) - (A + B)| = |a_n - A + b_n - B| \leq |a_n - A| + |b_n - B| < \frac{\varepsilon}{2} + \frac{\varepsilon}{2} = \varepsilon.$$

Therefore $\lim_{n\to\infty}(a_n + b_n) = A + B$, by definition.

For the second result, without loss of generality, assume $\{a_n\}$ diverges and $\{b_n\}$ converges to B. Suppose toward a contradiction that the sequence $\{a_n + b_n\}$ converges. Observe that because $\{b_n\}$ converges to B, the sequence $\{-b_n\}$ converges to $-B$. Then because $(a_n + b_n) + (-b_n) = a_n$, we have that $\{a_n\}$ converges, by the first result proved in this challenge. This contradicts the divergence of $\{a_n\}$. Thus our second result holds.

Remark: Students may need to be reminded of the power of the triangle inequality. Encourage them also to look for clever ways to add and subtract useful quantities within an absolute value in order to apply the triangle inequality in a productive way. Such a technique is used for Challenges 16.4 and 16.6.

16.4.   The statement is a theorem. *Extension:* For $\{a_n\}$ and $\{b_n\}$ satisfying the given conditions, $\lim_{n\to\infty} a_n = B$ if and only if $\lim_{n\to\infty} b_n = B$.

Proof of original statement: Observe that the condition on $\{a_n\}$ and $\{b_n\}$ implies that the absolute value of the difference between a_n and b_n is at most $1/n$. Now suppose $\{b_n\}$ converges to B and let $\varepsilon > 0$ be given. We know there exists an integer M such that for all $m \geq M$, $|b_m - B| < \varepsilon/2$. Let N be an integer such that $N > M$ and $1/N < \varepsilon/2$. Then for all $n \geq N$, we have $1/n < \varepsilon/2$. By the triangle inequality, we have

$$|a_n - B| = |a_n - b_n + b_n - B| \leq |a_n - b_n| + |b_n - B| < 1/n + \varepsilon/2 < \varepsilon/2 + \varepsilon/2 = \varepsilon.$$

So by definition, $\{a_n\}$ converges to B.

16.5.   The statement is false. *Counterexample:* Let $a_n = 1$ for all n. Then $\lim_{n\to\infty} a_n = 1$ because $|a_n - 1| = 0 < \varepsilon$ for all $\varepsilon > 0$ and all n. However, $\{b_n\}$ diverges because the terms alternate between 0 and 1 and thus cannot be made arbitrarily close to a limit value L. *Salvage:* The result holds if we redefine b_n as follows:

$$b_n = \begin{cases} a_n, & \text{if } n \text{ is even;} \\ A, & \text{if } n \text{ is odd.} \end{cases}$$

Proof: Let $\varepsilon > 0$ be given. Then there is an integer N so that for all $n \geq N$, we have $|a_n - A| < \varepsilon$. We also know that $|b_n - A| = |a_n - A|$ for even n and $|b_n - A| = |A - A| = 0$ for odd n; therefore $|b_n - A| < \varepsilon$ for all $n \geq N$. Thus $\lim_{n \to \infty} b_n = A$.

16.6.   *The statement is a theorem. Extension:* The sequence $\{a_n\}$ converges if and only if $\{a_n\}$ is a Cauchy sequence.

Proof of original statement: Suppose $\{a_n\}$ converges to A and let $\varepsilon > 0$ be given. Then there exists an integer N such that for all $n \geq N$, we have $|a_n - A| < \varepsilon/2$. Therefore for all m, n such that $n \geq m \geq N$, we apply our good friend the triangle inequality to obtain

$$|a_n - a_m| = |a_n - A + A - a_m| \leq |a_n - A| + |A - a_m| < \frac{\varepsilon}{2} + \frac{\varepsilon}{2} = \varepsilon.$$

Therefore $\{a_n\}$ is a Cauchy sequence.

Remark: To prove that every Cauchy sequence converges requires some fundamental ideas from real analysis. In fact, the property that all Cauchy sequences converge is equivalent to a property of the real numbers called *completeness*. Here is an opportunity to acknowledge to students just how complex the set of real numbers is. You might also point out that, in fact, one of the main goals of real analysis is to construct the real numbers using either Cauchy sequences or another technique called Dedekind cuts.

   Once students know that a Cauchy sequence converges, they will appreciate the idea that proving a sequence is Cauchy establishes its convergence without having to know its actual limit.

16.7.   *The statement is a theorem. Extension:* Let $\{a_n\}$ and $\{b_n\}$ be sequences satisfying $a_n \leq b_n$ for all n. Let $I_n = [a_n, b_n]$ and let k be a constant with $0 < k < 1$. If for all n, $I_{n+1} \subseteq I_n$ and $\text{length}(I_{n+1}) = k \times \text{length}(I_n)$, then $\{a_n\}$ and $\{b_n\}$ are both Cauchy sequences.

Proof of original statement: We have two infinite sequences $\{a_n\}$ and $\{b_n\}$ satisfying $a_n \leq b_n$ for all $n = 1, 2, 3 \ldots$, with I_n denoting the closed interval $[a_n, b_n]$. Because we know that $I_{n+1} \subseteq I_n$, we must have $a_n \leq a_m$ and $b_m \leq b_n$ for all $n \leq m$. Therefore, given $n \leq m$, we know that

$$b_n - a_n = (b_n - b_m) + (b_m - a_m) + (a_m - a_n),$$

the sum of three nonnegative terms. We also know that $\text{length}(I_{n+1}) = \frac{1}{2}\text{length}(I_n)$. Therefore we can inductively show that $\text{length}(I_n) = (\frac{1}{2^{n-1}})\text{length}(I_1) = \frac{b_1 - a_1}{2^{n-1}}$. Now let $\varepsilon > 0$ be given. Choose N sufficiently large so that $\frac{b_1 - a_1}{2^{N-1}} < \varepsilon$. Then for all $m \geq n \geq N$, we have $\text{length}(I_n) < \varepsilon$. Thus we also have

$$\text{length}(I_n) = b_n - a_n = (b_n - b_m) + (b_m - a_m) + (a_m - a_n) < \varepsilon.$$

From this inequality, we can conclude that each of the nonnegative terms in the sum above must also be less than ε. Therefore for all $m \geq n \geq N$, we have $|b_n - b_m| < \varepsilon$ and $|a_n - a_m| < \varepsilon$, which imply that both $\{a_n\}$ and $\{b_n\}$ are Cauchy sequences.

Remark: Note that this proof requires only slight modification to prove the extension.

16.8. To produce the *n*th stage, the number of subintervals removed from the $(n-1)$th stage is 2^{n-1}.

Proof: The number of subintervals removed to produce the *n*th stage is equal to the number of subintervals that exist at the $(n-1)$th stage. The first stage has 2 subintervals, the second stage has 4, and inductively we see that the $(n-1)$th stage has 2^{n-1} subintervals. The result follows.

16.9. The length of all removed intervals at the *n*th stage is $1-\left(\frac{2}{3}\right)^n$.

Proof: Let a_k be the length of one of the subintervals removed to produce the *k*th stage in the construction of the Cantor set. Then $a_1 = 1/3$ and $a_2 = 1/9$; so inductively, we see that $a_k = 1/3^k$. Applying Challenge 16.8, we deduce that the total length of subintervals removed to produce the *k*th stage is $\left(2^{k-1}\right)\left(\frac{1}{3^k}\right) = \frac{1}{3}\left(\frac{2}{3}\right)^{k-1}$. Therefore the length of all removed intervals at the *n*th stage is

$$\sum_{k=1}^{n}\frac{1}{3}\left(\frac{2}{3}\right)^{k-1} = \frac{1}{3}\sum_{k=0}^{n-1}\left(\frac{2}{3}\right)^k = \frac{1}{3}\left(\frac{1-\left(\frac{2}{3}\right)^n}{1-\frac{2}{3}}\right) = 1-\left(\frac{2}{3}\right)^n.$$

16.10. The statement is a theorem. *Extension:* See *Stepping back* at the end of the module.

Proof: The result of Challenge 16.9 implies that the lengths of all intervals that make up the set $[0,1] \setminus C$ equals $\lim_{n\to\infty}\left(1-\left(\frac{2}{3}\right)^n\right) = 1$.

Remark: This result implies something truly remarkable about the Cantor set: There is no piece of it that has positive length. In real analysis, such a set is said to have measure zero. Yet we will show in Challenge 16.14 that the Cantor set is uncountable. An uncountable set of measure zero is quite amazing for a bunch of points between 0 and 1!

16.11. The statement is false. *Counterexample:* The expression will never be negative and so cannot be used to express $\alpha < 0$. *Salvage:* Given any real number $\alpha \geq 0$, there exists an integer $N \geq 0$ such that

$$\alpha = \sum_{n=-N}^{\infty}\frac{a_n}{3^n}, \text{where each } a_n \in \{0, 1, 2\}.$$

Proof: The result is clear for $\alpha = 0$. If α is a positive integer, it can be written in base 3 by following the same process used to write an integer in base 10, as follows: Locate α on the number line between consecutive powers of 3—$3^N \leq \alpha < 3^{N+1}$. Subtract 3^N from α and repeat the process for the integer $3^N - \alpha$. We inductively obtain α as a sum of powers of 3.

So now we need only consider α with $0 < \alpha < 1$. We first locate α in the 1st, 2nd, or 3rd third of $[0, 1]$ and assign a value to a_1 as follows:

$$a_1 = \begin{cases} 0, & \text{if } 0 < \alpha < 1/3; \\ 1, & \text{if } 1/3 \leq \alpha < 2/3; \\ 2, & \text{if } 2/3 \leq \alpha < 1. \end{cases}$$

To determine the value of a_2, we locate α in the 1st, 2nd, or 3rd third of the subinterval found above. Using inequalities analogous to those in the definition of a_1, we have

$$a_2 = \begin{cases} 0, & \text{if } \alpha \text{ is in the 1st third of the subinterval;} \\ 1, & \text{if } \alpha \text{ is in the 2nd third of the subinterval;} \\ 2, & \text{if } \alpha \text{ is in the 3rd third of the subinterval.} \end{cases}$$

We continue this process. If at any stage we find α equals an endpoint of one of the thirds, then all subsequent values of a_i will equal 0, and we can stop. Otherwise we repeat the process to determine values for all a_n, $n = 1, 2, 3 \ldots$.

Remark: This important representation leads to the proof of Challenge 16.13, which is critical in the proof that the Cantor set is uncountable (Challenge 16.14).

16.12. The statement is false. *Counterexample:* The sums are not unique. For example, $0.111 = 0.1 + 0.011 = 0.01 + 0.101$. *Salvage:* Every real number in $[0, 1]$ can be expressed as the sum of two numbers whose ternary expansions are of the form

$$\sum_{n=0}^{\infty} \frac{a_n}{3^n}, \text{ where each } a_n \in \{0, 1\}.$$

Proof: Let r be any number in $[0, 1]$. From Challenge 16.11, we know that there are integers $a_n \in \{0, 1, 2,\}$ such that

$$r = \sum_{n=0}^{\infty} \frac{a_n}{3^n}.$$

Using this expansion, we will create two numbers, x and y, whose ternary expansions have the desired form. We will let b_n denote the coefficients for x and c_n denote the coefficients for y. Define the values of b_n and c_n as follows:

$$b_n = \begin{cases} 1, & \text{if } a_n = 1 \text{ or } 2; \\ 0, & \text{if } a_n = 0; \end{cases} \qquad c_n = \begin{cases} 1, & \text{if } a_n = 2; \\ 0, & \text{if } a_n = 0 \text{ or } 1. \end{cases}$$

Thus we have $a_n = b_n + c_n$ and, therefore, $r = x + y$.

16.13. The statement is a theorem. There is no natural extension.

Proof: Let c be a point in the Cantor set. By Challenge 16.11, we know that c has a ternary expansion

$$c = \sum_{n=0}^{\infty} \frac{a_n}{3^n}, \text{ where each } a_n \in \{0, 1, 2\}.$$

Notice that in determining the values of a_n in the proof of Challenge 16.11, we set $a_n = 1$ if and only if c lies in the 2nd (middle) third of the interval under consideration. Because $c \in C$, it will never lie in a middle third. Thus $a_n = 0$ or 2 for all $n = 1, 2, 3 \ldots$, and our result holds.

16.14.   The statement is false. *Salvage:* The Cantor set is not a countable set.

Proof: Challenge 16.13 implies that c is in $\mathcal{C}$ if and only if

$$c = \sum_{n=0}^{\infty} \frac{a_n}{3^n}, \text{ where each } a_n \in \{0, 2\}.$$

Thus the points in $\mathcal{C}$ are in one-to-one correspondence with the set of infinite sequences of 0's and 2's. Challenge 7.7 implies that the set of all infinite vectors of 0's and 1's is uncountable, so clearly the set of all infinite sequences of 0's and 2's is uncountable. Therefore $\mathcal{C}$ is uncountable.

16.15.   The statement is a theorem.

Proof: Take a real number $\alpha \in [0, 2]$. We divide by 2, obtaining $\frac{\alpha}{2} \in [0, 1]$. By 16.12, we know that

$$\frac{\alpha}{2} = \sum_{n=1}^{\infty} \frac{a_n}{3^n} + \sum_{n=1}^{\infty} \frac{b_n}{3^n}, \text{ where } a_n, b_n \in \{0, 1\}.$$

If we multiply both sides by 2, we get

$$\alpha = \sum_{n=1}^{\infty} \frac{a_n}{3^n} + \sum_{n=1}^{\infty} \frac{b_n}{3^n}, \text{ where } a_n, b_n \in \{0, 2\}.$$

Thus, both $\sum_{n=1}^{\infty} \frac{a_n}{3^n}$ and $\sum_{n=1}^{\infty} \frac{b_n}{3^n}$ are in the Cantor set.

Stepping back

We can modify the construction of the Cantor set to create a set that will have positive length. Begin with $[0, 1]$, which we denote $\mathcal{K}_0$. At stage 1, remove the middle half to create $\mathcal{K}_1$. At stage 2, create $\mathcal{K}_2$ by removing the middle quarter from each of the two intervals in $\mathcal{K}_1$. At stage 3, create $\mathcal{K}_3$ by removing the middle eighth from each of the four intervals in $\mathcal{K}_2$. In general, create $\mathcal{K}_n$ by removing the middle $\left(\frac{1}{2^n}\right)$th from each interval in $\mathcal{K}_{n-1}$. Thus, in particular, at each stage, the total length of $\mathcal{K}_{n-1}$ is reduced by a factor of $\frac{1}{2^n}$. Let $\mathcal{K}$ be the set obtained after all stages $n = 1, 2, 3, \ldots$ have been completed. We note that $\mathcal{K}$ has a structure analogous to the Cantor set. We will now show that the "length," or *measure,* of all the intervals in $\mathcal{K}$ is at least 1/4.

Let $m(\mathcal{K}_n)$ denote the measure of $\mathcal{K}_n$. So the measure of $\mathcal{K}$ will equal $m(\mathcal{K}) = \lim_{n \to \infty} m(\mathcal{K}_n)$. We know that $m(\mathcal{K}_0) = 1$. By its construction, we also know that for all $n = 1, 2, 3, \ldots$,

$$m(\mathcal{K}_n) = m(\mathcal{K}_{n-1}) - \frac{1}{2^n}(m(\mathcal{K}_{n-1})) = \left(1 - \frac{1}{2^n}\right)(m(\mathcal{K}_{n-1})).$$

We iterate this formula to obtain

$$m(\mathcal{K}_n) = \prod_{i=1}^{n}\left(1 - \frac{1}{2^i}\right)$$

and thus

$$m(\mathcal{K}) = \lim_{n \to \infty} \prod_{i=1}^{n} \left(1 - \frac{1}{2^i}\right), \text{ for } n = 1, 2, 3, \dots .$$

Now we show that each $m(\mathcal{K}_n) > \frac{1}{4}$ by manipulating the product formula. We illustrate our technique with an example for $n = 5$. We use the fact that the geometric series $\frac{1}{16} + \frac{1}{32} + \frac{1}{64} + \dots$ sums to $\frac{1}{8}$.

$$m(\mathcal{K}_5) = \prod_{i=1}^{5} \left(1 - \frac{1}{2^i}\right) = \frac{1}{2} \cdot \frac{3}{4} \cdot \frac{7}{8} \cdot \frac{15}{16} \cdot \frac{31}{32} = \frac{1}{2} \cdot \frac{3}{4} \cdot \frac{7}{8} \cdot \frac{15}{16} \cdot \frac{32-1}{32}$$

$$= \frac{1}{2} \cdot \frac{3}{4} \cdot \frac{7}{8} \cdot \frac{15}{16} \cdot \frac{32}{32} - \frac{1}{2} \cdot \frac{3}{4} \cdot \frac{7}{8} \cdot \frac{15}{16} \cdot \frac{1}{32}$$

$$= \frac{1}{2} \cdot \frac{3}{4} \cdot \frac{7}{8} \cdot \frac{16}{16} \cdot \frac{32}{32} - \frac{1}{2} \cdot \frac{3}{4} \cdot \frac{7}{8} \cdot \frac{1}{16} - \frac{1}{2} \cdot \frac{3}{4} \cdot \frac{7}{8} \cdot \frac{15}{16} \cdot \frac{1}{32}$$

$$= \frac{1}{2} \cdot \frac{3}{4} \cdot \frac{8}{8} \cdot \frac{16}{16} \cdot \frac{32}{32} - \frac{1}{2} \cdot \frac{3}{4} \cdot \frac{1}{8} - \frac{1}{2} \cdot \frac{3}{4} \cdot \frac{7}{8} \cdot \frac{1}{16} - \frac{1}{2} \cdot \frac{3}{4} \cdot \frac{7}{8} \cdot \frac{15}{16} \cdot \frac{1}{32}$$

$$> \frac{1}{2} \cdot \frac{3}{4} - \frac{1}{16} - \frac{1}{32} - \frac{1}{64} = \frac{1}{2} \cdot \frac{3}{4} - \left(\frac{1}{16} + \frac{1}{32} + \frac{1}{64}\right) > \frac{3}{8} - \frac{1}{8} = \frac{1}{4}.$$

We generalize this approach using induction to obtain

$$m(\mathcal{K}_n) = \prod_{i=1}^{n} \left(1 - \frac{1}{2^i}\right) > \frac{3}{8} - \left(\frac{1}{16} + \frac{1}{32} + \dots + \frac{1}{2^{n+1}}\right) > \frac{1}{4}.$$

Because each $m(\mathcal{K}_n) > \frac{1}{4}$, we have $m(\mathcal{K}_n) = \lim_{n \to \infty} m(\mathcal{K}_n) \geq \frac{1}{4}$.

Thus we have constructed a Cantor-like set with positive measure.

17

Uninterrupted thoughts of continuity
A jump-free journey

Building on the results of Module 16, this module introduces continuous functions. The definition for functions from $\mathbb{R}$ to $\mathbb{R}$ is given formally, and Challenges 17.1 through 17.5 cover some basic properties. Results on limits from Challenges 16.1 through 16.7 are used throughout the module. Encourage students to reconcile the δ, ε definition of continuity with the intuitive grasp they may already have about the concept: A continuous function is one for which the graph can be drawn without lifting the pencil from the paper. Or, a continuous function $f(x)$ is one for which, basically, very small changes in x result in reasonably small changes in $f(x)$.

There are a number of places in which we finesse some analytic rigor. The major occurrences are mentioned below; others are acknowledged within the solutions or in remarks. We remind instructors that Modules 16 and 17 foreshadow and highlight the basic ideas of real analysis and are not to replace a real analysis course. Thus we unabashedly assert that we will not give all the minute details. We acknowledge this reality because some instructors may wish to guide their students to the appropriate level of rigor they desire.

Challenge 17.6 (Intermediate Value Theorem) uses Challenge 16.7 and the fact that Cauchy sequences converge, as pointed out in the remark after Challenge 16.6 in this resource. Acknowledge to students that this sophisticated proof requires some results from analysis, offering them a peek at the many subtleties of analysis proofs.

Several challenges extend the idea of continuity to functions with domain other than $\mathbb{R}$. Challenges 17.7 and 17.8 consider continuous functions on the circle and the sphere, respectively. Challenges 17.6 and 17.10 through 17.15 consider functions that are assumed to be continuous on a closed interval. In all these cases, the text does not give formal definitions of these extended notions of continuity. This is another opportunity to acknowledge the many details that would be handled more rigorously in a full course on analysis. Point out to students that, for the purposes of this module, the intuitive ideas of a continuous function discussed above will serve them well in understanding continuous functions defined on a circle or a sphere. Continuity at the endpoints of a closed interval is another matter entirely, though you may wish to discuss it if your students are familiar with one-sided limits.

This material is well-suited for an introductory course on proof or a senior capstone course.

Solutions

17.1. The statement is a theorem. *Extension:* For real numbers a_0, ... , a_n, the function $f(x) = a_n x^n + a^{n-1} + \ldots + a_1 x + a_0$ is continuous.

Proof of original statement: If $m = 0$, $\ell(x) = b$ is a constant function. So for any $\varepsilon > 0$, we have $|\ell(x) - \ell(x_0)| = |b - b| = 0 < \varepsilon$ for all x and x_0. Thus $\ell(x)$ is continuous.

Suppose $m \neq 0$. Let x_0 be an arbitrary fixed real number and let $\varepsilon > 0$ be given. We need to find a $\delta > 0$ so that for every $x \in \mathbb{R}$ with $|x - x_0| < \delta$, it follows that $|f(x) - f(x_0)| < \varepsilon$. We work backward from what we want. First observe that $|f(x) - f(x_0)| = |(mx + b) - (mx_0 + b)| = |m(x - x_0)|$. Thus to obtain $|f(x) - f(x_0)| < \varepsilon$, we need $|m(x - x_0)| = |m||(x - x_0)| < \varepsilon$. So we choose $\delta = \varepsilon/|m|$, and the result follows.

Remark: As for the extension, many students will remember from calculus the fact that all polynomials are continuous. Challenge 17.9 requires the continuity of polynomials. Here's an outline of one way to prove it: First prove that the sum of any two continuous functions is continuous (Challenge 17.3). Then prove that the product of two continuous functions is continuous. We know from Challenge 17.1 that constant functions are continuous and that the function $f(x) = x$ is too; therefore, we can build up any polynomial we want as the sum and product of continuous functions.

17.2. The statement is false. *Counterexample:* Let $h(x)$ be a discontinuous function and let $\tau = 0$. Then $\tau h(x) = 0$ for all x and is continuous at all points. *Salvage:* If there exists a real number $\tau \neq 0$ such that the function $\tau h(x)$ is continuous, then $h(x)$ is a continuous function.

Proof: To show h is continuous, consider an arbitrary fixed real number x_0 and $\varepsilon > 0$. Note that $|\tau|\varepsilon > 0$, because $\tau \neq 0$. The function τh is continuous; so there is a $\delta > 0$ such that for every real x with $|x - x_0| < \delta$, we have $|\tau h(x) - \tau h(x_0)| < |\tau|\varepsilon$. Thus we also have $|\tau h(x) - \tau h(x_0)| = |\tau||h(x) - h(x_0)| < |\tau|\varepsilon$. Therefore $|h(x) - h(x_0)| < \varepsilon$, and h must be continuous.

Remark: This result implies that constant multiples of continuous functions are also continuous. Thus, in particular, if f is continuous, then so is $-f$.

17.3. The statement is a theorem. *Extension:* If $f_1(x), f_2(x), \ldots , f_n(x)$ are continuous functions for some positive integer n, then the function $f_1(x) + f_2(x) + \cdots + f_n(x)$ is continuous.

Proof of original statement: To show $f + g$ is continuous, consider an arbitrary fixed real number x_0 and $\varepsilon > 0$. Note that $\varepsilon/2 > 0$. The function f is continuous; so there is a $\delta_1 > 0$ such that for every real x with $|x - x_0| < \delta_1$, we have $|f(x) - f(x_0)| < \varepsilon/2$. The function g is continuous; so there is a $\delta_2 > 0$ such that for every real x with $|x - x_0| < \delta_2$, we have $|g(x) - g(x_0)| < \varepsilon/2$. Let δ be the smaller of δ_1 and δ_2. Then for every real x with $|x - x_0| < \delta$, by the triangle inequality, we also have $|(f + g)(x) - (f + g)(x_0)| = |f(x) - f(x_0) + g(x) - g(x_0)| \leq |f(x) - f(x_0)| + |g(x) - g(x_0)| < \varepsilon/2 + \varepsilon/2 = \varepsilon$. Therefore $f + g$ is also continuous.

Remark: The extension is easily proved using induction. Note also that by the remark following Challenge 17.2, if f and g are continuous, then so is $f - g$. We will use this fact in Challenge 17.7.

17.4.   The statement is false. *Counterexample:* Let $g(x) = -f(x)$. Then $f(x) - g(x) = 0$ for all x and is continuous, regardless of whether f or g is continuous. *Salvage:* For functions $f(x)$ and $g(x)$, if one is continuous at x_0 and the other is discontinuous at x_0, then $f(x) + g(x)$ is discontinuous at x_0.

Proof: Without loss of generality, suppose $f(x)$ is continuous at x_0 and $g(x)$ is not continuous at x_0. From the remark following Challenge 17.2, we know that $-f(x)$ is also continuous at x_0. Now suppose $f(x) + g(x)$ is continuous at x_0. Then by Challenge 17.3, we have $-f(x) + (f(x) + g(x)) = g(x)$ continuous at x_0, which is a contradiction. Therefore $f(x) + g(x)$ is not continuous at x_0.

17.5.   The statement is a theorem. *Extension:* See *Remark.*

Proof: We want to show that the sequence $\{f(x_n)\}$ converges to the number $f(L)$. Let $\varepsilon > 0$ be given. Because f is continuous, it is continuous at L. Thus for our given ε, there is a $\delta > 0$ such that whenever $|x - L| < \delta$, we have $|f(x) - f(L)| < \varepsilon$. In addition, because $\{x_n\}$ converges to L and $\delta > 0$, there exists an integer N such that for all $n \geq N$, we have $|x_n - L| < \delta$. Therefore for all $n \geq N$, we have $|f(x_n) - f(L)| < \varepsilon$. Thus our result holds.

Remark: Notice that we only used the fact that the sequence $\{x_n\}$ converges to L, not that it was Cauchy. Thus the result holds for all converging sequences. As we remarked after Challenge 16.6, a sequence converges if and only if it is Cauchy. Thus the original statement of the challenge does encompass all converging sequences. Though proving the convergence of Cauchy sequences is beyond the scope of this course, we will use this fact in the proof of the next challenge.

17.6.   *Proof:* We will find the desired x by setting up a sequence of nested intervals that "zoom in" on x. Challenges 16.7 and 17.5 will be very useful, as will the convergence of any Cauchy sequence, as stated in the remark for Challenge 17.5.

Following the hint in Appendix 1, let $a_1 = a$ and $b_1 = b$, and let m_1 denote the midpoint of $I_1 = [a_1, b_1]$. If $f(m_1) = y$, then the proof is complete. If $f(m_1) < y$, then let $a_2 = m_1$, $b_2 = b_1$, and $I_2 = [a_2, b_2]$. If $f(m_1) > y$, then let $a_2 = a_1$, $b_2 = m_1$, and $I_2 = [a_2, b_2]$.

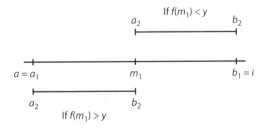

Schematic of possible intervals at first iteration

Now let m_2 denote the midpoint of I_2 and repeat the process. If $f(m_2) = y$, then the proof is complete. If $f(m_2) < y$, then let $a_3 = m_2$, $b_3 = b_2$, and $I_3 = [a_3, b_3]$. If $f(m_2) > y$, then let $a_3 = a_2$, $b_3 = m_2$, and $I_3 = [a_3, b_3]$. We iterate this process, at each stage creating a new interval $I_n = [a_n, b_n]$ half the length of I_{n-1}, with the additional property that $f(a_n) \leq y \leq f(b_n)$. The endpoints of the intervals form two infinite sequences, $\{a_n\}$ and $\{b_n\}$, with the property that $b_n - a_n = \frac{b-a}{2^n}$.

By Challenge 16.7, we know that $\{a_n\}$ and $\{b_n\}$ are Cauchy sequences. By the extension and remark following Challenge 16.6, we know that these sequences converge. Let $\lim_{n \to \infty} a_n = A$ and $\lim_{n \to \infty} b_n = B$. Note that the way our intervals were constructed implies that $a \leq A \leq B \leq b$. By Challenge 16.3, the difference $\{b_n - a_n\}$ also converges, and because $b_n - a_n = \frac{b-a}{2^n}$, we must have $\lim_{n \to \infty}(b_n - a_n) = 0$. Therefore $\lim_{n \to \infty} b_n - \lim_{n \to \infty} a_n = A - B = 0$, and we have $A = B$.

Now we apply Challenge 17.5. Because $\{a_n\}$ is a Cauchy sequence converging to A and because f is continuous, $f(a_n)$ converges to $f(A)$. Similarly, $\{b_n\}$ is a Cauchy sequence converging to B, so $f(b_n)$ converges to $f(B)$. Because we know that $f(a_n) \leq y \leq f(b_n)$ for $n = 1, 2, \ldots$, we must have $f(A) \leq y \leq f(B)$. But we also know that $A = B$. Therefore $y = f(A)$, with $a \leq A \leq b$, and the Intermediate Value Theorem holds.

Remark: Note that the theorem also holds if $f(b) < f(a)$. Many students will recall the Intermediate Value Theorem as a result that can be understood immediately from a sketch:

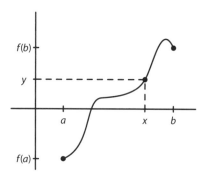

17.7.   The statement is a theorem. *Extension:* Let $\mathcal{K}$ be a sphere in $\mathbb{R}^3$, and let $F : \mathcal{K} \to \mathbb{R}$ be a continuous function. Then there exist two antipodal points X_1 and X_2 on $\mathcal{K}$ such that $F(X_1) = F(X_2)$.

Proof of original statement: We want to use the Intermediate Value Theorem to prove this result. First we have to define a function $g : \mathbb{R} \to \mathbb{R}$ that agrees with F as follows. Locate the circle $\mathcal{K}$ with its center at the origin. For every real number θ, the angle measured θ radians counterclockwise from the positive x-axis determines a point P on $\mathcal{K}$. Note that pairs of antipodal points on $\mathcal{K}$ correspond to pairs of angles θ and $\pi + \theta$.

Define $g : \mathbb{R} \to \mathbb{R}$ by $g(\theta) = F(P)$. Because F is continuous on $\mathcal{K}$, g is continuous on $\mathbb{R}$. Note also that $g(\theta) = g(\theta \pm 2\pi)$. Now define the function $f : \mathbb{R} \to \mathbb{R}$ by $f(\theta) = g(\theta) - g(\pi + \theta)$. If $f(\theta) = 0$ for all θ, then $g(\theta)$ is a constant, as is $F(P)$, and the result holds.

So we know there is some a for which $f(a) \neq 0$. Suppose for the first case that $f(a) < 0$. Then $g(a) - g(\pi + a) < 0$, so $f(\pi + a) = g(\pi + a) - g(2\pi + a) = g(\pi + a) - g(a) > 0$. Let $b = \pi + a$. So we have $f(a) < 0$ and $f(b) > 0$. By the remark following Challenge 17.3, f is continuous because g is continuous. In particular, f is continuous on the interval $[a, b]$. Now we apply the Intermediate Value Theorem (Challenge 17.6). Because $f(a) < 0 < f(b)$, there is a $c \in [a, b]$ such that $f(c) = 0$. So $g(c) = g(\pi + c)$. Let X_1 and X_2 be the points on $\mathcal{K}$ corresponding to c and $\pi + c$, respectively. Then $F(X_1) = g(c) = g(\pi + c) = F(X_2)$, so we have our antipodal points for which $F(X_1) = F(X_2)$.

For the second case, we have $f(a) > 0$. Relabel a with b and let $a = \pi - b$. Following the argument above will complete the proof.

Proof of extension: This proof is given in the next challenge.

17.8.   The answer is "Yes." There must always exist two antipodal points on Earth having the exact same temperatures.

Proof: We model Earth's surface with a sphere and recall some facts about the geometry of a sphere. Antipodal points are those pairs of points that are as far apart as possible. A great circle on a sphere is a circle with radius equal to the radius of the sphere; it is the intersection of the sphere with a plane through the center. In particular, each point on a great circle has its antipodal point on the same great circle.

Consider a fixed moment in time and define a function T on all points on Earth's surface as follows: For each point P, let $T(P)$ equal the temperature at P at that fixed moment in time. Clearly points very close to each other will have similar temperatures, so the function T is continuous. We now consider a great circle $\mathcal{C}$ on Earth's surface. Circle $\mathcal{C}$ lies in a plane, and the function T is continuous on $\mathcal{C}$. Thus by Challenge 17.7, there are two antipodal points on $\mathcal{C}$ with the same value of T. These points are antipodal on Earth's surface, so the result holds.

17.9.   The statement is false. *Counterexample:* Consider the function $P(x) = x^2 + 1$. At no real value of x does $P(x) = 0$. *Salvage:* Let $P(x)$ be a polynomial of odd degree. Then there exists a real number x_0 such that $P(x_0) = 0$.

Proof: We assume that polynomials are continuous. (See the remark following Challenge 17.1). Let $P(x) = a_n x^n + a_{n-1} x^{n-1} + \cdots + a_1 x + a_0$. Then $P(x) = a_n x^n \left(1 + \frac{a_{n-1}}{x} + \cdots + \frac{a_1}{x^{n-1}} + \frac{a_0}{x^n} \right)$. So for large positive or negative values of x, the value of $P(x)$ is dominated by the highest degree term. That is, if $a_n > 0$, then as x increases without bound, so does $P(x)$. Thus for sufficiently large x, we have $P(x) > 0$. Because n is odd, we also know that as x decreases without bound, so does $P(x)$. Thus for sufficiently negative x, we have $P(x) < 0$. The opposite behavior occurs in the case in which $a_n < 0$.

Thus P is a continuous function that takes on both positive and negative values. By the Intermediate Value Theorem, there is some x_0 for which $P(x_0) = 0$. Thus all polynomials of odd degree have at least one real root.

17.10.   We present the requested values with two claims.

Claim: $D(0) = 0$ and $D(1) = 1$

Proof: The ternary expansion of 0 is $0 = 0.00\ldots$, so we have $D(0) = 0.00\ldots$ in binary, which also equals 0. We claim that one way to expand the number 1 in ternary is $1 = 0.222\ldots$. To prove this, let $x = 0.222\ldots$ and consider the number $3x$. Just as multiplying a decimal expansion by 10 moves the decimal point one digit to the right, multiplying a ternary expansion by 3 moves the "decimal" point one digit to the right. So $3x = 2.222\ldots$. Thus $3x - x = 2$, which gives us $x = 1$. So now we can compute $D(1)$ to be $0.111\ldots$ in binary. We can pull the same trick on this expansion: Let $y = 0.111\ldots$. Then because our expansions are in binary, $2y = 1.111\ldots$, so $2y - y = 1$. Thus $y = 1$, and we have $D(1) = 1$.

This example shows that ternary and binary expansions are not necessarily unique, but $D(x)$ is still well-defined. The trick of moving the "decimal" point will reconcile the value of $D(x)$ for any x with a ternary expansion that ends in an infinite sequence of 2's.

Claim: For $x \in \left[\frac{1}{3}, \frac{2}{3} \right]$, $D(x) = \frac{1}{2}$. For $x \in \left[\frac{1}{9}, \frac{2}{9} \right]$, $D(x) = \frac{1}{4}$. For $x \in \left[\frac{7}{9}, \frac{8}{9} \right]$, $D(x) = \frac{3}{4}$.

Proof: For $x \in \left[\frac{1}{3}, \frac{2}{3} \right]$, a ternary expansion of x begins $x = 0.1\ldots$. Thus in the computation of $D(x)$, we have $N = 1$. So $D(x) = 0.1$ in binary, which equals $\frac{1}{2}$.

For $x \in \left[\frac{1}{9}, \frac{2}{9} \right]$, a ternary expansion of x begins $x = 0.01\ldots$. So we have $N = 2$, $b_1 = 0$, $b_2 = 1$. Thus $D(x) = 0.01$ in binary, which equals $\frac{1}{4}$.

For $x \in \left[\frac{7}{9}, \frac{8}{9} \right]$, a ternary expansion of x begins $x = 0.21\ldots$. So we have $N = 2$, $b_1 = 1$, $b_2 = 1$. Thus $D(x) = 0.11$ in binary, which equals $\frac{3}{4}$

17.11.   Here's a rough sketch of $y = D(x)$.

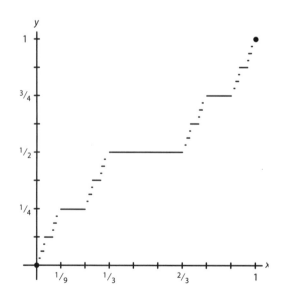

17.12. *Proof:* Consider $x_1, x_2 \in [0, 1]$ with $x_1 < x_2$. Write each number in ternary: $x_1 = 0.c_1c_2c_3 \ldots$ and $x_2 = 0.d_1d_2d_3 \ldots$. If these digits match up to and including the first occurrence of the digit 1, then by the definition of D, $D(x_1) = D(x_2)$. Otherwise, let k be the index of the first digit in which x_1 and x_2 differ. So $c_k < d_k$, and all earlier digits match and equal 0 or 2. There are two cases to consider: (1) Suppose $c_k = 0$ and $d_k = 1$. Then the kth digit of $D(x_1)$ will be 0 and the kth digit of $D(x_2)$ will be 1, with all earlier digits matching. Thus $D(x_1) < D(x_2)$. (2) Suppose $c_k = 1$ and $d_k = 2$. Then the kth digit of $D(x_1)$ will be 1, with all subsequent digits equal to 0. The kth digit of $D(x_2)$ will also be 1. Thus $D(x_1) \leq D(x_2)$.

17.13. *Claim:* The function D is locally constant for all points in the open intervals of middle thirds of $[0, 1]$. More precisely, let $x \in [0, 1]$ have ternary expansion $\sum_{n=1}^{\infty} \frac{a_n}{3^n}$, in which $a_n = 1$ for some $n = N$. Suppose further that for some $n > N$, $a_n \neq 0$. Then D is locally constant at x.

Proof: Let $x \in [0, 1]$ be as specified above. Let N be the smallest integer for which $a_N = 1$ and let $M > N$ be an integer such that $a_M \neq 0$. Then x lies in the open middle third with left endpoint $A = \sum_{n=1}^{N} \frac{a_n}{3^n}$. All the points in this open interval have ternary expansions that match x through a_N, which equals 1. By the proof of Challenge 17.12, all these points have function value $D(x)$. So D is locally constant on the open middle third containing x and thus on all open middle thirds.

Remark: We also claim that D is continuous on $[0,1]$. Being locally constant on all middle thirds implies D is continuous there as well. To show D is continuous at the remaining points (the Cantor set), requires looking carefully at the definition of D and the behavior of ternary expansions. We omit the details.

17.14. Let $S = \{x \in [0, 1] : D'(x) = 0\}$. The sum of the lengths of all intervals in S is 1.

Proof: For simplicity in this proof, we assume familiarity with the derivative. By Challenge 17.13, we know that D is locally constant on all open intervals of middle thirds in $[0,1]$. Thus on each of those intervals, $D'(x) = 0$. We also observe that for real numbers $a < b$, the open interval (a, b) has length $b - a$. Thus the sum of the lengths of all intervals in S equals the sum of the lengths of all middle thirds in $[0, 1]$. This is the same as the measure of $[0, 1] \setminus C$ for the Cantor set C, which Challenge 16.10 proved to be 1. The result follows.

17.15. The statement is false. *Counterexample:* See *Salvage.*

Salvage: The function $D : [0, 1] \to \mathbb{R}$ is an increasing continuous function such that $D(0) = 0$ and $D(1) = 1$. Yet there is no nontrivial interval I in $[0, 1]$ for which $D'(x) > 0$ for all $x \in I$.

Proof: Suppose there is a nontrivial interval I in $[0, 1]$ for which $D'(x) > 0$ for all $x \in I$. Let I have length $a > 0$. Then the lengths of all the intervals in $[0, 1]$ for which $D'(x) = 0$ sum to at most $1 - a$, which is less than 1. This contradicts the conclusion of Challenge 17.14, so the result holds.

Stepping back

We exhibit a sequence of continuous functions for which the limit function is not continuous. Let $f_n(x) = x^n$ for $n = 1, 2, \ldots$. Then for each n, f_n is continuous and maps $[0, 1]$ into $[0, 1]$. For each x_0 with $0 \le x_0 < 1$, we have $f(x_0) = \lim_{n \to \infty}(x_0)^n = 0$. Also, $f(1) = \lim_{n \to \infty} 1^n = 1$. Therefore the limit function f also maps $[0, 1]$ into $[0, 1]$ but is clearly not continuous at $x = 1$.

Remark: This challenge demonstrates that when a sequence of continuous functions converges pointwise, the limit function need not be continuous. We need a stronger condition than pointwise convergence. Let $\{f_n\}$ be a sequence of continuous functions from a set S to $\mathbb{R}$. We say that the sequence $\{f_n\}$ *converges uniformly* to the function f if for any given $\varepsilon > 0$, there is an integer N such that for all indices $n \ge N$, it follows that $|f_n(x) - f(x)| < \varepsilon$ for *all* $x \in S$. Note that given $\varepsilon > 0$, a uniformly converging sequence of functions can use the same δ for all values of x in the domain. This stronger characteristic will guarantee that the limit function of a uniformly converging sequence of functions is also continuous.

18

An abstract world of algebra

Reconciling with your ✗

This module introduces abstract algebra through elementary group theory. The challenges develop fundamental results and basic properties, including those of abelian groups, direct products, and group tables. Challenge 18.5 introduces $\mathbb{Z}_n$ and thus refers to material from Module 4 on modular arithmetic. Challenge 18.8 relies on Fermat's Little Theorem (Challenge 4.10). The *Stepping back* challenge foreshadows the idea of isomorphism. All challenges are very accessible, despite the fact that they introduce some classic proving techniques in abstract algebra. The material is ideally suited for an introduction to proof course, a pre-teacher course, or a capstone course.

Solutions

18.1. The solutions below combine parts (a) and (b) for each operation $*$. Student answers to part (b) will depend on how they chose to salvage flawed operations. In most cases, the claims follow immediately from the properties of well-known arithmetic operations. For parts (v) and (vi), brief arguments are given.

(i) $* : \mathbb{Z} \times \mathbb{Z} \to \mathbb{Z}$ defined by $a * b = a + b$ is a binary operation with identity element $e = 0$.

(ii) $* : \mathbb{Z}^+ \times \mathbb{Z}^+ \to \mathbb{Z}^+$ defined by $a * b = a - b$ is not a binary operation. When $b \geq a$, the value $a * b$ does not lie in $\mathbb{Z}^+$. *Salvage:* Replace each $\mathbb{Z}^+$ with $\mathbb{Z}$. The resulting binary operation has no identity element.

(iii) $* : \mathbb{Q} \times \mathbb{Q} \to \mathbb{Q}$ defined by $a * b = ab$ is a binary operation with identity element $e = 1$.

(iv) $* : \mathbb{Q} \times \mathbb{Q} \to \mathbb{Q}$ defined by $a * b = a/b$ is not a binary operation. When $b = 0$, the value $a * b$ is not defined. *Salvage:* Replace each $\mathbb{Q}$ with $\mathbb{Q} \setminus \{0\}$. The resulting binary operation has no identity element.

(v) $* : \mathbb{R} \times \mathbb{R} \to \mathbb{R}$ defined by $a * b = \sqrt{a + b}$ is not a binary operation. When $a + b < 0$, the value $a * b$ is not defined. *Salvage:* Replace each $\mathbb{R}$ with $\mathbb{R}^+$. The resulting binary operation does not have an identity element. There is no positive real number e with the property that for all $a \in \mathbb{R}^+$, $\sqrt{a + e} = a$.

(vi) $*: \mathbb{R} \times \mathbb{R} \to \mathbb{R}$ defined by $a * b = \sqrt{a^2 + b^2}$ is a binary operation. It does not have an identity element because, for example, $a * b$ is never negative. Thus for any $a < 0$, there is no $e \in \mathbb{R}$ for which $a * e = a$.

(vii) $*: \mathbb{C} \times \mathbb{C} \to \mathbb{C}$ defined by $a * b = \pm\sqrt{a + b}$ is not a binary operation. It is not even a function, because $\pm\sqrt{a + b}$ is not well-defined. There is no way to change the set $\mathbb{C}$ to correct this flaw.

18.2.   The statement is a theorem. *Extension:* $\langle \mathbb{Q}, + \rangle$, $\langle \mathbb{R}, + \rangle$ are also groups.

Proof of original statement: We know from Challenge 18.1 that addition is a binary operation on $\mathbb{Z}$ and that $e = 0$ is an identity element. We know from deep in our mathematical past that addition is associative. We also know that given $a \in \mathbb{Z}$, if we let $a' = -a$, then we have $a + (-a) = (-a) + a = 0$. Thus every element has an inverse, and $\langle \mathbb{Z}, + \rangle$ is a group.

Remark: This argument works to prove the extension.

18.3.   The statement is a theorem. *Extension:* If $\langle G, * \rangle$ is a group, then there is exactly one identity element.

Proof of extension: Because G is a group, we know it has at least one identity element. We need to show the identity element is unique. Suppose e_1 and e_2 are both identity elements in G. We will show that $e_1 = e_2$. Because e_1 is an identity, we have $e_2 * e_1 = e_1 * e_2 = e_2$. But e_2 is also an identity; so we have $e_1 * e_2 = e_2 * e_1 = e_1$. Therefore, $e_1 = e_2$.

Remark: This is a classic uniqueness proof. Encourage students to remember this approach, especially for Challenge 18.4. To show some element is unique, consider two such elements and show they must be equal. Point out that students do not need to include the assumption that the two elements are distinct. Doing so will lead to an unnecessary proof by contradiction: assuming two distinct elements, showing they are equal, reaching a contradiction. Note that within that three-step argument lies a direct proof that the elements are equal.

18.4.   The statement is a theorem. *Extension.* Let $\langle G, * \rangle$ be a group. If $a, b \in G$, then the equation $a * x = b$ has a unique solution $x \in G$ and the equation $y * a = b$ has a unique solution $y \in G$.

Proof of extension: First we find a solution to the equation $a * x = b$. We want $b = a * x$, so we compute

$$
\begin{aligned}
a' * b &= a' * (a * x) &&\text{(where } a' \text{ is the inverse of } a\text{)} \\
&= (a' * a) * x &&\text{(by the associativity of } *\text{)} \\
&= e * x &&\text{(by a property of inverses)} \\
&= x. &&\text{(by a property of the identity)}
\end{aligned}
$$

Substituting $x = a' * b$ into the equation $a * x = b$, we apply group properties again to find

$$a * x = a * (a' * b) = (a * a') * b = e * b = b,$$

as desired. Therefore $x = a' * b$ is a solution to the equation.

Now suppose $w \in G$ is also a solution; so $a * x = a * w$. Thus $(a' * a) * x = (a' * a) * w$, which implies that $x = w$. Following an analogous argument, we find the unique solution to $y * a = b$ is $y = b * a'$.

Remark: Note that the uniqueness result of this challenge embodies a *cancellation property* for groups: $a * x = a * w$ if and only if $x = w$. Students might also recognize that this challenge implies that the inverse of a group element is unique. Given $a \in G$, we know that $a * a' = a' * a = e$; therefore a' is the unique element that satisfies the two equations $a * x = e$ and $x * a = e$. This observation is very helpful in Challenges 18.6 and 18.7. Note also that these same observations imply $(a')' = a$. We will use this fact in Challenge 19.1.

18.5.   The statement is false. *Counterexample:* Look at the case when $n = 4$. Under addition modulo 4, $\mathbb{Z}$ has no identity. For example, $4 \oplus 0 \equiv 0 \bmod 4$, but we want $4 \oplus 0 \equiv 4$. A similar conflict arises for each n. Thus $\mathbb{Z}$ is never a group under addition modulo n. *Salvage:* Let $n > 1$ be a fixed integer, let $\mathbb{Z}_n = \{0, 1, \ldots, n - 1\}$, and let $\oplus$ denote addition modulo n. Then $\langle \mathbb{Z}_n, \oplus \rangle$ is a group.

Proof: For $a, b \in \mathbb{Z}_n$, $a \oplus b \in \mathbb{Z}_n$ by the definition of addition modulo n. The operation $\oplus$ is also associative, because ordinary addition is associative. The identity of $\mathbb{Z}_n$ is 0 because $a \oplus 0 = 0 \oplus a = a$ for all $a \in \mathbb{Z}_n$. To establish inverses, first note that 0 is its own inverse. For $a \in \mathbb{Z}_n$, $a \neq 0$, observe that because $0 < a < n$, we also have $0 < n - a < n$. Thus we have $a' = n - a$, because $a + (n - a) = n$ is equivalent to 0 mod n.

18.6.   The statement is a theorem. *Extension:* The definition can be extended to any finite product of groups: $\langle G_1 \times G_2 \times \cdots \times G_n, * \rangle$ is a group for any natural number n, where each G_i is a group and $*$ is defined component-wise.

Proof of original statement: Let $G = \langle G_1 \times G_2, * \rangle$. Because G_1 and G_2 are groups, the operation defined on G is a binary, associative operation in each component. If $e_1 \in G_1$ and $e_2 \in G_2$ are identity elements, then for $(a_1, a_2) \in G$ we have

$$(a_1, a_2) * (e_1, e_2) = (a_1 *_1 e_1, a_2 *_2 e_2) = (a_1, a_2) \text{ and } (e_1, e_2) * (a_1, a_2) = (e_1 *_1 a_1, e_2 *_2 a_2) = (a_1, a_2).$$

From the extension of Challenge 18.3, we know the identity is unique. Therefore (e_1, e_2) is the identity in G. Finally, because G_1 and G_2 are groups, for $(a_1, a_2) \in G$ we have inverses $a_1' \in G_1$ and $a_2' \in G_2$. Thus $(a_1', a_2') \in G$, and we have

$$(a_1, a_2) * (a_1', a_2') = (a_1 *_1 a_1', a_2 *_2 a_2') = (e_1, e_2) \text{ and } (a_1', a_2') * (a_1, a_2) = (a_1' *_1 a_1, a_2' *_2 a_2) = (e_1, e_2).$$

From the remark following Challenge 18.4, we know inverses are unique. Therefore $(a_1, a_2)' = (a_1',a_2')$, and G is a group.

Remark: The extension is easily shown by induction.

18.7.   The statement is a theorem. *Extension:* The result holds for all groups. If G is a group, then for all $a, b \in G$, $(a * b)' = b' * a'$.

Proof of extension: From the remark following Challenge 18.4, we know the inverse of $a * b$ is the unique element x that satisfies both $(a * b) * x = e$ and $x * (a * b) = e$. Thus once we show that $b' * a'$ also satisfies the two equations, we will know $(a * b)' = b' * a'$. Observe:

$$
\begin{aligned}
(a * b) * (b' * a') &= (a * (b * b')) * a' && \text{(by associativity)} \\
&= (a * e) * a' && \text{(by property of inverses)} \\
&= a * a' && \text{(by property of identity)} \\
&= e. && \text{(by property of inverses)}
\end{aligned}
$$

Similarly we have

$$(a' * b') * (b * a) = (a' * (b' * b)) * a = (a' * e) * a = a' * a = e.$$

Therefore $(a * b)' = b' * a'$.

Remark: Students may notice that this result can be extended to products of more than two elements. Thus for $a_1, a_2, \dots, a_n \in G$, $(a_1 a_2 \cdots a_n)' = a_n' \cdots a_2'a_1'$. This extension is shown easily by induction.

18.8.   The statement is false. *Counterexample:* In $\mathbb{Z}_4$, 2 has no inverse. *Salvage:* Let $\mathbb{Z}_n^*$ be the nonzero elements of $\mathbb{Z}_n$ where n is prime. Then $\mathbb{Z}_n^*$ is a group under multiplication modulo n.

Proof: We observe that $1 \in \mathbb{Z}_n^*$. Because elements of $\mathbb{Z}_n^*$ are all less than n and n is prime, multiplication mod n will never yield a product of 0 and, therefore, defines an associative binary operation on $\mathbb{Z}_n^*$ with identity 1.

Let $a \in \mathbb{Z}_n^*$. To show that a has an inverse, we call upon Fermat's Little Theorem (Challenge 4.10). By definition of $\mathbb{Z}_n^*$, every $a \in \mathbb{Z}_n^*$ is relatively prime to n. So by Fermat's Little Theorem, we have $a^{n-1} \equiv 1$ mod n. Thus a^{n-2} will be the inverse of a in $\mathbb{Z}_n^*$, and our proof is complete.

18.9.   The group tables for $\langle \mathbb{Z}_4, \oplus_4 \rangle$ and $\langle \mathbb{Z}_7, \oplus_7 \rangle$ are given below. The group table for an abelian group must be symmetric about the main diagonal. That is, the mth row must be identical to the mth column for $m = 1, 2, \dots, |G|$.

$\mathbb{Z}_4$	0	1	2	3
0	0	1	2	3
1	1	2	3	0
2	2	3	0	1
3	3	0	1	2

$\mathbb{Z}_7$	0	1	2	3	4	5	6
0	0	1	2	3	4	5	6
1	1	2	3	4	5	6	0
2	2	3	4	5	6	0	1
3	3	4	5	6	0	1	2
4	4	5	6	0	1	2	3
5	5	6	0	1	2	3	4
6	6	0	1	2	3	4	5

18.10. *Claim:* Let $\langle G, * \rangle$ be a finite group. Then each row (column) of the group table for G contains each element of G exactly once.

Proof: Let $G = \{g_1, g_2, \dots, g_n\}$, where $n = |G|$. We proceed by contradiction and assume that row m of the group table has element g_j appearing twice. Then there exist distinct elements g_i and g_k such that $g_m * g_i = g_j$ and $g_m * g_k = g_j$. This violates the extension to Challenge 18.4, which guarantees a unique solution to $g_m * x = g_j$. Thus no row of the group table can contain repeated elements. A similar argument implies the result for columns.

Stepping back

As discussed in the hint, this challenge leads students to discover the idea of isomorphism. To better reflect the approach that students might use for this challenge, this solution assumes no familiarity with the definition or proof techniques of isomorphism, but it does comment on the concept as it arises naturally.

Claim: There are two possible groups with four elements, as shown in the tables.

G_1	g_1	g_2	g_3	g_4
g_1	g_1	g_2	g_3	g_4
g_2	g_2	g_3	g_4	g_1
g_3	g_3	g_4	g_1	g_2
g_4	g_4	g_1	g_2	g_3

G_2	g_1	g_2	g_3	g_4
g_1	g_1	g_2	g_3	g_4
g_2	g_2	g_1	g_4	g_3
g_3	g_3	g_4	g_1	g_2
g_4	g_4	g_3	g_2	g_1

Proof: We need to verify two claims: First, each of the tables above represents a group. Second, there are no other "distinct" groups with four elements. For the first task, we compare each table above with the group table for a known group. Compare the table for G_1 with the table for $\mathbb{Z}_4$, shown in the solution to Challenge 18.9. Relabel the elements of G_1 as follows: Replace g_1 with 0, replace g_2 with 1, replace g_3 with 2, and replace g_4 with 3. The group table for G_1 is then transformed into exactly the table for $\mathbb{Z}_4$. Thus because $\mathbb{Z}_4$ is a group, we also know that G_1 is a group under the operation defined by the table above.

For G_2, consider the table below for the group $\mathbb{Z}_2 \oplus \mathbb{Z}_2$.

$\mathbb{Z}_2 \oplus \mathbb{Z}_2$	$(0,0)$	$(1,0)$	$(0,1)$	$(1,1)$
$(0,0)$	$(0,0)$	$(1,0)$	$(0,1)$	$(1,1)$
$(1,0)$	$(1,0)$	$(0,0)$	$(1,1)$	$(0,1)$
$(0,1)$	$(0,1)$	$(1,1)$	$(0,0)$	$(1,0)$
$(1,1)$	$(1,1)$	$(0,1)$	$(1,0)$	$(0,0)$

Relabel the elements of G_2 as follows: Replace g_1 with (0,0), replace g_2 with (1,0), replace g_3 with (0,1), and replace g_4 with (1,1). The group table for G_2 is then transformed into exactly the table for $\mathbb{Z}_2 \oplus \mathbb{Z}_2$. Thus because $\mathbb{Z}_2 \oplus \mathbb{Z}_2$ is a group, we also know that G_2 is a group under the operation defined by the table above.

The process we just used to show that G_1 and G_2 are groups illustrates the idea of two groups being *isomorphic*. Informally we say G_1 is isomorphic to $\mathbb{Z}_4$, because both groups have essentially the same algebraic structure but different names for the elements. This idea is fundamental in abstract algebra. There are analogous concepts in every field of mathematics. One of the basic tasks in mathematics is determining when two structures, such as groups, are really the same, even when the elements and operations might look different.

Now we need to show that G_1 and G_2 are the only possible groups with four elements, up to isomorphism. We first note that every group must have an identity, which we label g_1 without loss of generality. Every element must have an inverse, and we already know that $g_1' = g_1$; so for now suppose $g_2' = g_4$. Thus $g_2 g_4 = g_4 g_2 = g_1$, which implies $g_2 = g_4'$. So we have that g_2 and g_4 are inverses of each other. Because inverses are unique, we must have $g_3' = g_3$. So far this is consistent with the table for G_1:

G_1	g_1	g_2	g_3	g_4
g_1	g_1	g_2	g_3	g_4
g_2	g_2	–	–	g_1
g_3	g_3	–	g_1	–
g_4	g_4	g_1	–	–

Challenge 18.10 implies that each row and each column must contain each group element once. So in the g_2 row, we must have $g_2 g_3$ equal to g_4, which forces $g_2 g_2$ to equal g_3. Having completed the second row of the table, we now apply Challenge 18.10 repeatedly to complete the table. The result is exactly the table for G_1. Had we supposed that $g_2' = g_3$, we would have obtained a structurally identical (isomorphic) group, with the labels g_3 and g_4 interchanged.

To obtain our second group, we again let g_1 be the identity, but we now consider the only remaining possibility for g_2'—that $g_2 g_2 = g_1$. If we suppose further that g_3 and g_4 are inverses of each other, then we will replicate the structure of G_1 again. The only other option is for g_3 and g_4 to be their own inverses. Thus we have a table consistent with G_2:

G_2	g_1	g_2	g_3	g_4
g_1	g_1	g_2	g_3	g_4
g_2	g_2	g_1	–	–
g_3	g_3	–	g_1	–
g_4	g_4	–	–	g_1

As with the first group, Challenge 18.10 forces the remaining entries in the table to match the table for G_2. Thus we have shown that any group with four elements must be structurally equivalent to either G_1 or G_2. More specifically, any group with four elements is isomorphic to either $\mathbb{Z}_4$ or $\mathbb{Z}_2 \oplus \mathbb{Z}_2$. Notice finally that every group with four elements is abelian.

19

Cycles and curves
Algebraic structure in numbers and geometry

Building on the material of Module 18, this module introduces cyclic groups, which connect some basic results in number theory with some basic structures in algebra. Challenges 19.1 through 19.7 explore fundamental results, using modular arithmetic from Module 4 for finite cyclic groups of integers. Challenge 19.6 uses the result of Challenge 3.10(b) on relatively prime numbers. Challenge 19.7 uses Challenge 3.4 (the Division Algorithm). Challenges 19.8 through 19.10 and *Stepping back* show that there is algebraic structure within geometric objects.

This material is well-suited for an introduction to proof course, a course for pre-teachers, or a senior capstone course.

Solutions

19.1. The statement is false. *Counterexample:* Let H be the empty set. Then H vacuously satisfies the given criterion, but H is not a subgroup because it does not contain the identity. *Salvage:* Let $\langle G, * \rangle$ be a group and H be a nonempty subset of G. Then $\langle H, * \rangle$ is a subgroup of G if and only if $a * b' \in H$ for all $a, b \in H$.

Proof: If $\langle H, * \rangle$ is a subgroup of G, then for all $a, b \in H$, we have $b' \in H$. We also know that $*$ is a binary operation on H. Thus $a * b' \in H$, and we are done.

For the (more interesting) converse, we assume that H is a nonempty subset of G satisfying the given criterion. To show H is a subgroup, we need to show that $*$ is an associative, binary operation on H; that H contains an identity; and that every element of H has an inverse in H.

Let e denote the identity of G. Because H is nonempty, we have some $a \in H$. Therefore by our condition, $a * a' = e \in H$. Because e acts as an identity for all elements of G, it is also an identity for H. If $b \in H$, then again by our condition, we have $e * b' = b' \in H$. Thus every element of H has an inverse in H.

The operation $*$ is associative for elements of H because it is associative on G. Therefore we need only show that $*$ is a binary operation on H (i.e., that $a * b \in H$ for all $a, b \in H$). We'll need the following fact, noted in the remark after Challenge 18.4: For any group element g, $(g')' = g$.

Consider $a, b \in H$. Then we know that b' is also in H. Therefore the given condition on H implies that $a * (b')' \in H$. Thus $a * b \in H$, and we have our binary operation. Therefore H is a subgroup.

Remark: This result gives a very compact criterion for when a subset of a group is a subgroup. We will use it in the next challenge. Notice also that in a subgroup, the identity and inverse elements are exactly the same as those in the parent group.

19.2. The statement is a theorem. *Extension:* The set $n\mathbb{Q} = \{a/b : a$ is a multiple of n, $\gcd(a, b) = 1$, $b \neq 0\}$ is a subgroup of $\mathbb{Q}$ under addition.

Proof of original statement: We will apply the result of Challenge 19.1. First we observe that $n\mathbb{Z}$ is nonempty. Now suppose $a, b \in n\mathbb{Z}$. Then $a = nt_1$ and $b = nt_2$ for integers t_1 and t_2. We know $b' = -nt_2$. Therefore $a + b' = nt_1 + (-nt_2) = n(t_1 - t_2)$. Because $t_1 - t_2$ is an integer, we have $a + b' \in n\mathbb{Z}$. Thus by Challenge 19.1, $n\mathbb{Z}$ is a subgroup of $\mathbb{Z}$.

19.3. The group $m\mathbb{Z}$ is a subgroup of $n\mathbb{Z}$ if and only if $n \mid m$. Thus subgroups of $\mathbb{Z}$ are connected with notions of divisibility.

Proof: By Challenge 19.2, we know that both $m\mathbb{Z}$ and $n\mathbb{Z}$ are subgroups of $\mathbb{Z}$ under addition. They share the operation and identity of $\mathbb{Z}$, and inverses are consistent with those of $\mathbb{Z}$. Thus $m\mathbb{Z}$ will be a subgroup of $n\mathbb{Z}$ if and only if $m\mathbb{Z}$ is a subset of $n\mathbb{Z}$.

Suppose that $m\mathbb{Z} \subseteq n\mathbb{Z}$. Then $mt \in n\mathbb{Z}$ for every integer t. In particular, we have $m \in n\mathbb{Z}$. Thus $m = nt$ for some t, which implies $n \mid m$.

Now suppose $n \mid m$. Then $m = nd$ for some integer d. Therefore every element mt of $m\mathbb{Z}$ can be written $mt = ndt$ and thus lies in $n\mathbb{Z}$. So $m\mathbb{Z} \subseteq n\mathbb{Z}$.

19.4. *Proof:* Consider $a \in \mathbb{Z}_n$. Then $a^n = a + a + \cdots + a$, n times. So $a^n = na$. Thus every element $m \in \mathbb{Z}_n$ can be written $m = m \cdot 1 = 1^m$. Therefore 1 is a generator for $\mathbb{Z}_n$.

Remark: This proof relates additive notation to multiplicative notation; the result $a^n = na$ can be confusing. Reassure students that multiplicative notation is the norm, with additive notation typically used only for groups where the operation is basically addition, such as $\langle \mathbb{Z}, + \rangle$. Multiplicative notation offers many conveniences: a' is denoted a^{-1} and all regular rules of exponents apply.

We also note that to be a generator of a group G with $|G| = n$, an element $a \in G$ must have the following property: The smallest positive integer k for which $a^k = e$ must be $k = |G|$. Then all the powers $a, a^2, a^3, \ldots, a^k = e$ are distinct and comprise all of G. If not, then $a^i = a^j$ for some $1 \leq i < j \leq k$, which yields $a^{j-i} = e$. But $j - i < k$, which contradicts the choice of k. This observation will be useful in Challenge 19.6.

19.5. The statement is false. *Counterexample:* $\mathbb{Z}_2 \times \mathbb{Z}_2$ is abelian but not cyclic. (See *Remark*.) *Salvage:* If $\langle G, * \rangle$ is a cyclic group, then it is an abelian group.

Proof: Because G is cyclic, it has a generator a. Given $g, h \in G$, we have $g = a^m$ and $h = a^n$ for integers m and n. Thus $g * h = a^m * a^n = a^{m+n} = a^{n+m} = a^n * a^m = h * g$, and G is abelian.

Remark: Encourage students to compute powers of each element in $\mathbb{Z}_2 \times \mathbb{Z}_2$ to convince themselves that the group has no generator. For example, $(1, 1)^2 = (1, 1) * (1, 1) = (1 \oplus_2 1, 1 \oplus_2 1) = (0, 0) = e$. Note that we can also think of this computation as $(1, 1)^2 = (2 \bmod 2, 2 \bmod 2) = (0, 0) = e$. The next challenge also requires students to be comfortable with computing powers of such elements.

19.6.   The group $G = \mathbb{Z}_4 \times \mathbb{Z}_2$ is not cyclic.

Proof: G has eight elements; therefore any generator a of G would have to have eight distinct powers: a, $a^2, a^3, \ldots, a^7, a^8 = a^0 = e$. But computation reveals that $a^4 = e$ for every $a \in G$, so there is no generator. For example, $(3, 1)^2 = (3, 1) * (3, 1) = (3 \oplus_4 3, 1 \oplus_2 1) = (2, 0)$. Therefore $(3, 1)^4 = (2, 0)^2 = (2 \oplus_4 2, 0)$ $= (0, 0) = e$. Note that we can also do this computation as $(3, 1)^4 = (12, 4) = (0, 0)$.

The group $G = \mathbb{Z}_4 \times \mathbb{Z}_3$ is cyclic with generator $(1, 1)$.

Proof: We compute the powers of $(1, 1)$ to show it generates all elements of G.

$$(1, 1)^2 = (1, 1) * (1, 1) = (1 \oplus_4 1, 1 \oplus_3 1) = (2, 2)$$
$$(1, 1)^3 = (3, 3) = (3, 0)$$
$$(1, 1)^4 = (4, 4) = (0, 1)$$
$$(1, 1)^5 = (5, 5) = (1, 2)$$
$$(1, 1)^6 = (6, 6) = (2, 0)$$
$$(1, 1)^7 = (7, 7) = (3, 1)$$
$$(1, 1)^8 = (8, 8) = (0, 2)$$
$$(1, 1)^9 = (9, 9) = (1, 0)$$
$$(1, 1)^{10} = (10, 10) = (2, 1)$$
$$(1, 1)^{11} = (11, 11) = (3, 2)$$
$$(1, 1)^{12} = (12, 12) = (0, 0) = e \ .$$

The group $G = \mathbb{Z}_m \times \mathbb{Z}_n$ is cyclic if and only if m and n are relatively prime.

Proof: We first note that G has mn elements. Suppose m and n are relatively prime. To show that G is cyclic, we will show that $(1, 1)$ is a generator—that is, we will show that the smallest k for which $(1, 1)^k = (0, 0)$ is $k = mn$. Recall the result of Challenge 3.10(b): If m and n are relatively prime and each divides an integer k, then $mn \mid k$. Observe that $(1, 1)^k = (k, k) = (0, 0)$ only if m and n both divide k. Thus we know that $mn \mid k$. We also know that $(1, 1)^{mn} = (mn, mn) = (0, 0)$. Therefore mn is the smallest power of $(1, 1)$ to equal $(0, 0)$. Therefore $(1, 1)$ is a generator for G, and G is cyclic.

For the converse, suppose m and n are not relatively prime. Then there is some integer $d > 1$ that divides both m and n. So mn/d is an integer divisible by both m and n. Observe that $(a, b)^{mn/d}$ $= (amn/d, bmn/d) = (0, 0)$. Thus for any $(a, b) \in G$, the smallest power of (a, b) giving $(0, 0)$ is less than mn. Therefore no element of G is a generator, and G is not cyclic.

19.7.   The statement is a theorem. *Extension:* Every subgroup of a cyclic group is cyclic.

Proof of extension: Suppose G is a cyclic group with generator a. Then $G = \{a^n : n \in \mathbb{Z}\}$. Let H be a subgroup of G, so that every element of H can be written as a^n for some integer n. If $H = \{e\}$, then H is generated by $a^0 = e$, and we are done. Otherwise H must contain at least one positive power of a. Let m be the smallest positive integer for which $a^m \in H$. We will show that a^m is a generator for H.

Let a^n be an arbitrary element of H. By the Division Algorithm (Challenge 3.4), there exist integers q and r for which $n = mq + r$, with $0 \leq r < m$. Thus $a^n = a^{mq+r} = a^{mq}a^r = (a^m)^q a^r$. Therefore $(a^m)^{-q}a^n = a^r$. We know that a^m is in H, and because H is a group, any power of a^m lies in H, as does the inverse of any power. Thus $(a^m)^{-q}$ lies in H. We also have a^n in H, and because H is a group, the product of any two elements also lies in H. Therefore $(a^m)^{-q}a^n = a^r$ lies in H. But $r < m$; so by the choice of m, we must have $r = 0$. Therefore $a^n = a^{mq}$, and a^m is a generator of H. Thus H is cyclic.

Remark: This elegant result is a testament to the power of the Division Algorithm. Encourage students to also think about how the proof suggests a fundamental quality of every cyclic group. The argument above considers products of elements of a group G. Because G is cyclic, all these products are powers of the generator. The powers are all integers, and thus the argument depends only on properties of ordinary arithmetic operations of integers, including the Division Algorithm. Students who construct a proof specifically for $\langle \mathbb{Z}, + \rangle$ might compare notes with those who prove the extension. Look for an opportunity to suggest that all cyclic groups are somehow structurally "the same" as cyclic groups of integers under addition. The idea of two groups being structurally the same, called *isomorphic*, was suggested in *Stepping back*, Module 18.

19.8.   *Proof:* Geometrically this claim is obvious: Given a circle C in the plane and a line ℓ through one point on the circle, ℓ will either be tangent to C or will intersect it in exactly one other point. We can also look at this result algebraically. The circle C is the set of points that satisfy a quadratic equation $(x - h)^2 + (y - k)^2 = r^2$; the line ℓ is the set of points that satisfy a linear equation $y = mx + b$ or, if ℓ is vertical, $x = c$.

The points of intersection will be those (x, y) satisfying both equations—that is, for which x satisfies $(x - h)^2 + (mx + b - k)^2 = r^2$ or, if ℓ is vertical, for which y satisfies $(c - h)^2 + (y - k)^2 = r^2$. Each of these are quadratic equations and therefore have two solutions. Because we know ℓ intersects C in at least one point, $\mathcal{O}$, we know that both solutions are real. If the solutions are distinct, then ℓ intersects C in a second point. If the solutions are not distinct, then ℓ is tangent to C at $\mathcal{O}$.

19.9.   *Proof:* Let $P \neq \mathcal{O}$ be a point on the circle. Let $L_{P\mathcal{O}}$ denote the line through P and $\mathcal{O}$. By definition, $P \oplus \mathcal{O}$ is the point where the line through $\mathcal{O}$ parallel to $L_{P\mathcal{O}}$ intersects the circle. In this case, the parallel line through $\mathcal{O}$ must be $L_{P\mathcal{O}}$ itself, which intersects the circle at P. Thus $P \oplus \mathcal{O} = P$. A similar argument shows $\mathcal{O} \oplus P = P$.

By the definition of P^{-1}, the line through $\mathcal{O}$ parallel to $L_{PP^{-1}}$ is tangent to the circle at $\mathcal{O}$. Thus $P \oplus P^{-1} = \mathcal{O}$. Similarly $P^{-1} \oplus P = \mathcal{O}$.

If $P = \mathcal{O}$, then $L_{P\mathcal{O}}$ is the line through $\mathcal{O}$ tangent to the circle (a line that intersects the circle only at $\mathcal{O}$). Thus $\mathcal{O} \oplus \mathcal{O} = \mathcal{O}$ and $\mathcal{O}' = \mathcal{O}$.

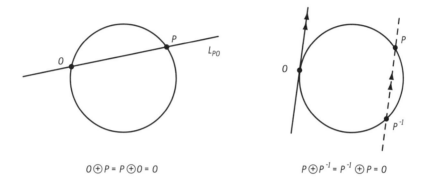

$\mathcal{O} \oplus P = P \oplus \mathcal{O} = \mathcal{O}$ $P \oplus P^{-1} = P^{-1} \oplus P = \mathcal{O}$

19.10.   The statement is a theorem. *Extension:* See *Stepping back.*

Proof of original statement: By definition, $\oplus$ is a binary operation on the points of $\mathcal{C}$. Challenge 19.9 implies the existence of an identity $\mathcal{O}$ and inverses for each point. The only property left to verify is that $\oplus$ is associative.

Following the hint in Appendix 1, let C denote the center of the circle and define the following: If P is a point on the circle, then the angle $\angle OCP$ is the angle determined by the points $\mathcal{O}$, C, and P, measured clockwise. Restricting t to the interval $[0, 2\pi)$, we have a one-to-one correspondence between points P on the circle and angles $t \in [0, 2\pi)$, measured clockwise from $\mathcal{O}$.

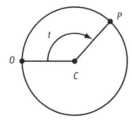

The angle $t = \angle OCP$

Now consider two points on the circle. Moving clockwise from $\mathcal{O}$, we label the points P and Q. Let $\alpha = \angle OCP$ and $\beta = \angle PCQ$. As illustrated below, we now claim that the point $P \oplus Q$ is the point corresponding to the angle $2\alpha + \beta$ modulo 2π. This claim can be verified using plane geometry, though we will not do so here. Thus we see that the associativity of $\oplus$ is equivalent to the associative property of addition mod 2π.

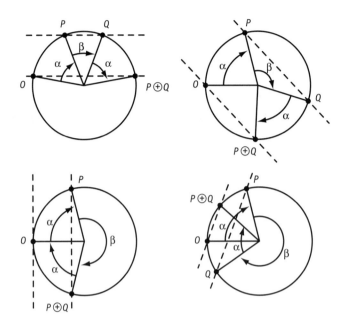

As an illustration, we show below on the left $(P \oplus Q) \oplus R$, with α and β the angles used in computing the first sum and α' and β' the angles used in computing the second sum. On the right, we show $P \oplus (Q \oplus R)$, with α and β the angles used in computing the first sum and α' and β' the angles used in computing the second sum.

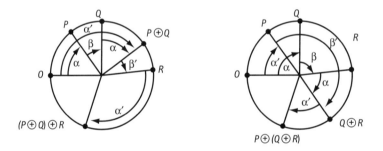

Stepping back

Let $\mathcal{C}$ be the graph of a conic section in the plane, and let $\mathcal{O}$ be a specified point on $\mathcal{C}$. We can define a binary operation $\oplus$ on the points of $\mathcal{C}$ just as $\oplus$ was defined on the points of a circle. For points P and Q on $\mathcal{C}$, the point $P \oplus Q$ is the point where the line ℓ through $\mathcal{O}$ parallel to L_{PQ} intersects $\mathcal{C}$. Once again, in the case that ℓ is tangent to $\mathcal{C}$ at $\mathcal{O}$, $P \oplus Q = \mathcal{O}$.

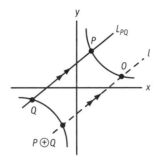

The operation $\oplus$ for the hyperbola $xy = 1$ with designated point $\mathcal{O}$

$\mathcal{C}$ is the set of points satisfying a quadratic equation, and the line ℓ parallel to L_{PQ} is the set of points satisfying a linear equation. As with the case of the circle, we know that ℓ intersects $\mathcal{C}$ in at least one point. Therefore $P \oplus Q$ is well-defined, and $\oplus$ is a binary operation on $\mathcal{C}$. All the group properties proved in Challenge 19.9 still hold. Proving associativity is tedious but doable. We leave the final details to those who have far too much time on their hands.

20

Further frontiers

The final module mirrors actual research in mathematics. Half of the ten challenges are open questions, but students don't know which ones. What a wonderful fantasy if a student were to crack an unsolved problem!

Some of the challenges require basic knowledge of prime numbers, infinite series (including the harmonic series), rational numbers, or the Pythagorean Theorem, as indicated in the solutions or comments that follow. A few require more detailed background: Challenge 20.5 builds on Module 16's work on the Cantor set; Challenge 20.6 applies to RSA coding introduced in Module 4; Challenge 20.10 extends work in Module 13 on the Art Gallery Theorem.

Solutions

20.1. The sequence converges to a sum less than 80.

Proof: We will bound the series with a converging geometric series. The bound may appear extremely crude at first, but there is a method to this madness.

First observe that

$$\frac{1}{1}+\frac{1}{2}+\frac{1}{4}+\cdots+\frac{1}{9}<1+1+\cdots+1=8.$$

Now observe that

$$\frac{1}{10}+\frac{1}{11}+\frac{1}{12}+\frac{1}{14}+\cdots+\frac{1}{19}<\frac{1}{10}+\frac{1}{10}+\cdots+\frac{1}{10}=\frac{9}{10}.$$

This inequality also holds for each sum with denominators in the 20's, 40's, and on through the 90's. There are eight such "decades," including the one above, for a total bound of $8 \cdot \frac{9}{10}$. Notice that we exclude the 30's.

Next observe that

$$\frac{1}{100}+\frac{1}{101}+\frac{1}{102}+\frac{1}{104}+\cdots+\frac{1}{109}<\frac{1}{100}+\frac{1}{100}+\cdots+\frac{1}{100}=\frac{9}{100}.$$

This inequality also holds for each sum with denominators in the 110's, 120's, 140's, and on through the 190's. There are nine such sums, including the one above, for a total bound of $9\cdot\frac{9}{100}=\left(\frac{9}{10}\right)^2$. We obtain the same bound for the sums with denominators in the 200's, 400's, and on through the 900's (skipping the 300's). There are a total of eight such "centuries," including the 100's, for a bound of $8\cdot\left(\frac{9}{10}\right)^2$.

We continue this analysis to obtain

$$\sum_{n\in\text{TF}}\frac{1}{n}<8+8\cdot\frac{9}{10}+8\cdot\left(\frac{9}{10}\right)^2\cdots=8\sum_{n=0}^{\infty}\left(\frac{9}{10}\right)^n=8\cdot\frac{1}{1-\frac{9}{10}}=80.$$

Thus the series converges, and its sum is less than 80.

This result may seem surprising, because we know the harmonic series diverges. However, when you think about a typical term, $\frac{1}{n}$, in the harmonic series, the number n is an integer with many digits. The chances that none of those digits are 3's is extremely small. In fact, as n gets very large, the chance that n has no 3's goes to 0. Thus most integers have been excluded from the set **TF**, so the sum $\sum_{n\in\text{TF}}\frac{1}{n}$ is almost like a finite sum.

20.2.   This is an open question known as the Goldbach Conjecture. Christian Goldbach, an 18th-century Prussian mathematician, conjectured that every even integer greater than 2 can be written as the sum of two primes. As of December 2005, computer searches have shown that the conjecture is true for all even $n\le 3\times 10^{17}$. Most mathematicians believe the conjecture is true based on statistical analyses, though such evidence does not constitute a proof.

20.3.   The answer is "No." To see this, suppose there exist positive rational numbers $r=a/b$ and $s=c/d$ for which $\sqrt[n]{r^n+s^n}$ is the rational number e/f. Then

$$\left(\frac{e}{f}\right)^n=\left(\frac{a}{b}\right)^n+\left(\frac{c}{d}\right)^n,$$

and so $(ebd)^n=(adf)^n+(cbf)^n$, which gives us a positive integer solution to $x^n+y^n=z^n$ for some $n>2$. But by Fermat's Last Theorem, no such solutions exist.

20.4.   If the product $\prod_{p\text{ a prime}}\left(1-\frac{1}{p}\right)^{-1}$ is expanded, we claim that it equals $\sum_{n=1}^{\infty}\frac{1}{n}$, the harmonic series. Because this series diverges, the product also diverges. Assuming this claim, we notice that if there were only finitely many primes, then the product would be a product of a finite number of rational numbers. This would imply that the harmonic series converges, which is a contradiction.

Now we establish the claim that $\prod_{p \text{ a prime}} \left(1 - \frac{1}{p}\right)^{-1} = \sum_{n=1}^{\infty} \frac{1}{n}$.

Proof: We rewrite the product to obtain

$$\prod_{p \text{ a prime}} \left(1 - \frac{1}{p}\right)^{-1} = \prod_{p \text{ a prime}} \frac{1}{1 - \frac{1}{p}}.$$

Because p is a prime, $\frac{1}{p} < 1$. So the expression $\frac{1}{1 - \frac{1}{p}}$ is the sum of a geometric series. Thus we have

$$\prod_{p \text{ a prime}} \frac{1}{1 - \frac{1}{p}} = \prod_{p \text{ a prime}} \left(\sum_{n=0}^{\infty} \left(\frac{1}{p}\right)^n \right) = \sum_{n=0}^{\infty} \left(\frac{1}{2}\right)^n \cdot \sum_{n=0}^{\infty} \left(\frac{1}{3}\right)^n \cdot \sum_{n=0}^{\infty} \left(\frac{1}{5}\right)^n \cdots$$

$$= \left(1 + \frac{1}{2} + \frac{1}{4} + \cdots\right)\left(1 + \frac{1}{3} + \frac{1}{9} + \cdots\right)\left(1 + \frac{1}{5} + \frac{1}{25} + \cdots\right) \cdots.$$

Notice that the reciprocal of every power of every prime appears exactly once in the last expression. Now imagine expanding that product of summations in a formal generalization of FOIL. Because every positive integer n is the unique product of a set of powers of primes, the reciprocal $\frac{1}{n}$ will appear exactly once in the expansion. For example, $\frac{1}{50} = \frac{1}{2} \cdot 1 \cdot \frac{1}{25} \cdot 1 \cdot 1 \cdots$. Thus we have

$$= \left(1 + \frac{1}{2} + \frac{1}{4} + \cdots\right)\left(1 + \frac{1}{3} + \frac{1}{9} + \cdots\right)\left(1 + \frac{1}{5} + \frac{1}{25} + \cdots\right) \cdots = \sum_{n=1}^{\infty} \frac{1}{n},$$

which establishes our claim.

20.5. It is possible to construct such a set. We now save several thousand words by displaying several pictures. (Recall how many words a picture is worth.)

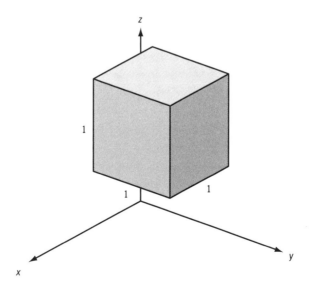

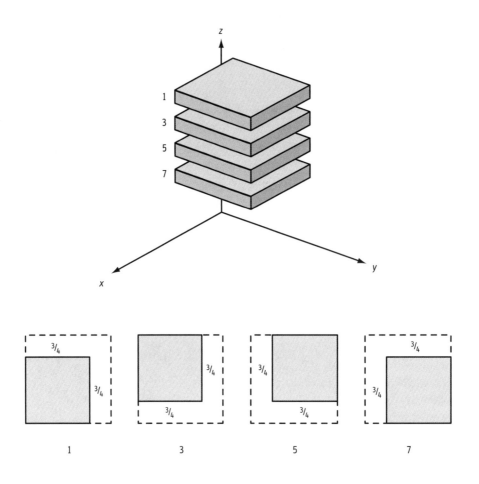

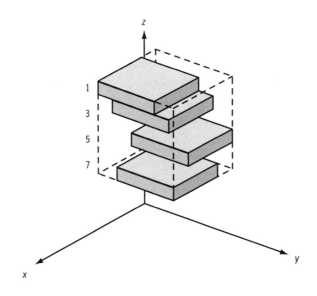

20.6.   This is an open question. If there were an easy way to break the code without factoring n, then the code scheme would be worthless. Because we don't know if such a method exists, for all practical purposes, it doesn't exist. Therefore, the RSA coding scheme is safe for the moment. Also, if factoring n could be done in polynomial time, RSA coding would be considerably less safe.

20.7.   This is an open question known as the $3x + 1$ problem. Using vast amounts of computer time, researchers have shown that for any starting number up to $426 \cdot 2^{50}$, the procedure always ends with a 1. An Internet search will reveal many resources on this problem. Students might also want to explore the analogous problem for $5x + 1$.

20.8.   This is an open question. Current work leads mathematicians to conjecture that there are no odd perfect numbers. Moreover, only a finite number of even perfect numbers are known. Finding even perfect numbers is related to finding prime numbers of the form 2^{n-1}, known as Mersenne primes. Marin Mersenne was a 17th-century monk who studied both prime numbers and perfect numbers.

20.9.   Yes, there are arbitrarily large gaps between consecutive primes. *Claim:* Given a positive integer N, there exist consecutive prime numbers p and q, $p < q$, such that $q - p \geq N$.

Proof: Recall that $N! = N(N-1)(N-2) \cdots 3 \cdot 2 \cdot 1$. Let p be the largest prime such that $p \leq N! + 1$, and let q be the smallest prime such that $N! + N + 1 \leq q$. Note that $p < q$ and $q - p \geq N! + N + 1 - (N! + 1) = N$. Now observe that k divides $N! + k$ for all integers k, $2 \leq k \leq N$. Thus none of the numbers $N! + 2$, $N! + 3, \ldots, N! + N$ can be prime. Therefore p and q are consecutive primes, and our claim holds.

20.10.   This is an open question. Ernst Straus posed the question in the 1950s. It is related to the Art Gallery Theorem from Module 13. A polygon for which the property holds is called illuminable. Conjectures suggest that all polygons are illuminable. Students might also wonder whether a polygon will always be illuminable from *each* point in its interior. The answer is "no," though counterexamples are complicated figures with many sides.

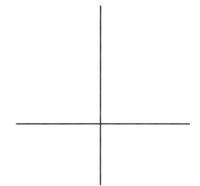

Appendix
A math personality questionnaire

Those readers who wish to collaborate with another person or groups of people might find it useful to answer the following questionnaire. It could help an instructor or group leader find individuals with compatible mathematical personalities, potentially leading to happy partnerships.

For each statement, please assign a number response from 0 to 10: 0 = disagree strongly; 5 = maybe/maybe not; 10 = agree strongly. **There are no correct answers!** Answer as honestly as you can.

1.   I'm an outgoing person.

2.   I am extremely competitive.

3.   I'm a bit insecure of my mathematical abilities.

4.   I hope my partner will not ask a bunch of silly questions.

5.   Math is not my number one priority.

6.   I like thinking about math by myself first before talking with others.

7.   I sometimes feel I work slower than others.

8.   I like chatting with friends about math homework and then thinking about the questions on my own.

9.   I prefer to work on my own.

10.   I like learning from others.

11.   A math partnership should be a give-and-take relationship.

12.   I wish I could do math all the time.

13.   I want to consider my partner as a resource.

14.   I hope I'll be able to interact with my partner on a regular basis.

15.   I love being right.

16.   I love learning something new from a classmate.

17.   I love making mistakes because I almost always gain some new insight.

18.   I am really excited about thinking about mathematics.

19.   I'd like my partner to be low-key.

20.   I like working on math after midnight.

21.   I like working on math at the last minute.

22.   Once I find a solution, I like to think more deeply about the issue.

23.   I plan to study more math after this course.

24.   I like to start on my homework as soon as possible.

25.   I think my mathematics professor is the cat's pajamas.